mySAP FI Fieldbook

mySAP FI Fieldbook

Thomas H. Spitters, CPA
Certified mySAP Accounting and
Business Integration Consultant

iUniverse, Inc.
New York Lincoln Shanghai

mySAP FI Fieldbook

iUniverse books may be ordered through booksellers or by contacting:

iUniverse
2021 Pine Lake Road, Suite 100
Lincoln, NE 68512
www.iuniverse.com
1-800-Authors (1-800-288-4677)

ISBN-13: 978-0-595-35018-6 (pbk)
ISBN-13: 978-0-595-79723-3 (ebk)
ISBN-10: 0-595-35018-6 (pbk)
ISBN-10: 0-595-79723-7 (ebk)

Printed in the United States of America

To Victoria

Contents

mySAP Financials—An Introductory Chapter 1

Financial Supply Chain Management and mySAP Accounting Tools 34

Overview of mySAP Reporting Tools 55

An Overview of mySAP Strategic Enterprise Management 72

An Afterword 104

Index 107

mySAP Financials—An Introductory Chapter

This chapter is intended for use as an informal introduction into mySAP Accounting and R/3 Enterprise Financials for the end-user. Some familiarity with SAP is assumed in this book. For instance, there is no discussion in this text of logging on or off the system, basic screen configuration in FI, SAP terms, navigation, business processes or why enterprises need integrated software. This chapter and book are ideal for the new project team member, financial manager or controller, or corporate staff in need of some immediate orientation about mySAP Accounting.

Many thanks to Jeffrey Dileo, who introduced me to SAP. Acknowledgment is also necessary to Jan Kohn, Jim Flanagan and John Kryczewski of SAP America, as well as Joan Black, Wayne Skitt and Steve Zablocki of SAP Canada.

An Overview of SAP Client/Server Architecture

A typical SAP Client/Server system is as follows: Hundreds of users are logged on to the various servers from all over the world running Sales and Distribution, Logistics, Financials and so on. There are a number of things that guarantee that a mySAP system runs and continues to run in this wide—reaching computing environment: high availability, high performance and data security. Applications are made available to end-users in different areas through LAN and WAN technology. A mySAP system performs in such a way as to keep all systems updated continuously, regardless of their location. The integrity and security of the entire system requires an administrative function made available through a help desk, and other system management and monitoring tools, which provide fast and efficient support for desktop problems that arise. Furthermore, network management in this type of computing environment is extremely important, especially concerning system performance and monitoring.

From a computing standpoint, the different applications in an SAP environment are concerned with process automation, workload distribution and balancing, performance optimization, and improving the capacity and availability of the servers to do their various planned computing tasks. Applications in SAP need to be managed in terms of the number of users and/or jobs on different servers, tools to diagnose problems as well as alerts in place to recognize or avoid problems including application to end user problems involving disk operation, monitoring and memory. The information gleaned from diagnostics in the system is important for capacity planning as well.

Server requests coming from the applications are handled by an underlying database that additionally supports tape drives, hard disks and networking facilities, etc., necessary to support SAP and a quality data processing environment. The performance of the database is measured through its availability and fault—tolerance which calls for protecting the data against unforeseen events and circumstances through the implementation of a mirrored database or a reliable backup and restore system. Additional integrity can be ensured by moving archive data offsite to a remote location, proper practices and procedures for data management and disaster recovery, proper attention to the operating system on top of the database layer including management tools that would indicate, for instance, available disk space and memory usage. With this in mind, in any database, the overall processing environment constantly changes, and performance optimization and tuning require frequent attention in an SAP system, especially during upgrades or data reorganizations, for instance. Through automating many of these tasks and duties, the SAP environment is designed to avoid computing problems and disasters in order to assure a high performing and high availability ERP system.

SAP Financials Overview

MySAP Financials are an integral part of SAP, a comprehensive business software package designed to operate in a variety of IT environments. SAP FI is part of the SAP core, that is, the basic enterprise resource planning software package offered by SAP, headquartered in Walldorf, Germany. Other elements of the SAP core include CO, SD, MM, AM, PP and HR. The focus of this publication will primarily be on the accounting function of SAP, however. For information on other modules, you are free to find information through your local bookseller, company library, or through SAP at www.saptech.com.

This publication will gloss over the integration points of FI with the other modules as the author wishes to maintain a strict accounting focus. FI is designed for legally required external reporting according to GAAP or the IAS framework. For the most part, reporting functionality down to the transaction level is available in SAP 4.6C or R/3 Enterprise Accounting, but there are also "bolt-on" applications, or third-party software available for unique reporting requirements such as sales tax in the U.S. FI is configurable in various languages, principally German and English, and is also usable in multilingual enterprises across international borders. SAP enables the translation of accounting information into multiple languages without a specific license, though in some portions of the module there are missing translations filled with German language terminology. The emphasis of R/3 4.6C was primarily a principle of portability of the information system across the world wide web and making the functionality of R/3 more easily operable to end users. R/3 Enterprise carries this a step further by not only emphasizing internet capability, but in its design and implementation making the intranet structure of the enterprise more important and more accessible to developers and end users, as well as emphasizing return on IT investment.

Functional applications of SAP FI include all the financial operations of an enterprise on the international level, the country—specific level, the consolidated entity level, and company, departmental area, and transaction levels. SAP Financials have the capability to deal with multiple currencies, entities, corporate structures, accounting and auditing information systems, charts of accounts and ledgers in a broad range of business activities in the same information system. The information system requires an operating system and a database, and SAP takes a holistic view of the enterprise to ensure integration of processes and data at all levels. SAP Financials have been developed with the consideration of customer requirements in a wide variety of economic sectors including banking, manufacturing, transportation and many others. SAP is designed to be non-industry specific, however, as the system is made up of components and processes basic to all enterprises.

SAP Financials has the capability, especially with the development of SEM and BW, to further optimize the accounting and financial function from data input to financial and management reporting. SAP Reporting Tools have developed into part of an enterprise solution in SAP BW that makes for a more quickly executed accounting cycle and better integration between accounting, financial operations and strategic management. This is due to an optimization of data loads, increases in software performance and the level of accounting detail, better integration with other modules, and better management tools.

SAP Financials also has seamless integration with Cost Center Accounting, Controlling, Asset Accounting, Travel Management, Consolidation, Investment and Strategic Enterprise Management applications within SAP, and with other third-party "bolt-on" applications. Financials can now communicate with SEM components such as Strategy Management, Performance Measurement, Business Consolidation, Business Planning and Stakeholder Relationship Management to optimize the relationship between the information system and business operations.

The real accounting and financial breakthrough of R/3 Enterprise is in the legal and management reporting realm. Reporting in R/3 Enterprise is significantly enhanced over 4.6C, more user—friendly and fully web—enabled. R/3 Enterprise reporting allows for multiple integration of data with scaled integration to other SAP applications. MySAP Financials now are structured to reduce the time of the accounting cycle, enhance end-user and management satisfaction, and to provide an optimal reporting solution. Results of implementation of SAP Financials in R/3 Enterprise include greater project management controls, faster financial closes, greater cost center management control, ability to globally assess profitability and financial measures, ability to globally assess operational costs, and a standardized system of system integration, training and support.

The Accounting Cycle in MySAP Financials

In mySAP Financials, the general ledger is the center of the accounting system, and all FI, CO, and asset management documents post to general ledger accounts or have related postings. Logistics modules (SD, MM, and PP) also have postings to the general ledger when recording receipts of materials, shipments or transfers of raw materials or finished goods. HR also has some limited integration with FI concerning accounting for labor and employee benefits at the end of each pay period.

The chart of accounts and master data files are common to all the modules integrating with FI: a. General ledger accounts are common to FI, SD, MM, PP, CO, HR and asset management; b. Customer master records are common to FI and SD; c. Vendor master records are common to FI, MM, and PP; d. Material master records for products are also found in SD, MM, PP, and WM. FI accounts are called "cost elements" in CO—CO is designed for internal reporting be it for cost centers (departments), profit centers or market segments.

FI is designed for legal accounting under GAAP and IAS for financial statements, VAT and required transaction journals. Sales tax computation, processing and reporting that is required only in the United States is handled through a third party software such as Vertex.

The customer master record, found in both FI and SD has screens ordinarily maintained through accounts receivable or order processing, customer service, and warehouse management. Multiple shipping addresses are possible for one bill—to address as a common customer database is available to all departments. You can update general ledger accounts through SD for customer shipments and returns, and maintenance of customer master records varies from organization to organization.

The vendor master records, found in MM and FI, can be maintained in accounts payable, purchasing, receiving, warehouse management or logistics departments. A vendor database is available to all these departments and it depends on your organization how the vendor master records are maintained. It is also possible to manage inventories through MM (measures of both quantity and value are available), and inventory stock movements update the general ledger. Maintenance of the material master records depends upon your company policy.

The Accounting Transaction Dialog Process in SAP

In an R/3 system the business application tasks are shared by the database, application server and presentation servers which allows for open distribution of transaction services over different computer systems. With the database server and application servers started, the transaction process begins with the dispatcher which starts up all the other work processes. The application server has its own memory management including a roll file and a local buffer mechanism for R/3 objects. Communication with the presentation server is managed by the dispatcher using a queue.

With the presentation server started, transaction processing options are available to the end user. The user executes a function from the menu or command line, and the presentation server passes the user request to the application server. As soon as it arrives the response time begins. The incoming request goes into a queue in the application server managed by the dispatcher. The time the user request must wait in the queue is called the wait time. User requests are processed one after the other. If a work process is available the dispatcher takes a user request from the queue and passes it to the free work process.

The application context and the transaction are loaded from the roll file to the work process. This process is called roll in. When the user request arrives in the work process, the roll time begins and the application server begins processing the request. This is when the dispatch time begins. SQL statements go from the ABAP interpreter to the database interface, also part of the work process. The database interface queries information from the SAP repository to check for data locally so data can be read from local, or main memory (this process takes place very quickly). As access to main memory is very fast, you must buffer SAP tables properly in order to ensure the proper rapid processing here.

After the database interface returns data to ABAP interpreter, processing of statements continues. Again, the ABAP interpreter recognizes the SQL statement and forwards it to DB interface—if a statement is not available locally it goes to the underlying DB directly—this is the start of database request time (this process is much slower than the process involving local memory). SQL statements must be optimized in order to assure proper speed of processing at the DB level.

After DB interface has received all requested data, the database request time ends and sends the data back to ABAP interpreter where processing continues. After processing of all statements has been completed, the work process goes to the next user screen. This is also the end of dispatch time—dispatch time indicates how long the request occupied the work process. The user context is also saved to the roll file (this process if referred to as roll out). The next user screen is sent back to the presentation server with screen data compressed, as typically only a small amount of data has to be sent to the presentation server. With this in mind, distributed computing is facilitated by the SAP dialog structure. When the screen is sent back to the presentation server, the response time ends. The presentation server prepares the screen for a graphical window environment and the new screen is finally presented to the end user.

When a transaction dialog process is finished, it prepares an update notification (this is a request at startup) for the update server, it prepares the data to be written to the update database tables. The update table reads the notification and processes the update. If a transaction has been set up for asynchronous updating, then the transaction does not carry out the database update itself. Rather, it passes the update to the update work processes. The update processes then carry out the database changes described in the update record. The transfer to the update work process takes place when the user transaction executes a database commit. The update data is passed to the update task in the form of a record which contains not only the data to be changed or updated but also instructions as to how to carry out the database change. That is, the update record describes: a. update modules to be called to execute the database change, and b. the data to be processed in a database update. The transaction that requests the update also creates a header for managing the update.

The update is performed dynamically depending upon which tables on which host are available for data. This dynamic process is brokered by the message server that maintains a list of processes and available hosts in the system and the list is updated continuously for changes in the system, i.e., the starting, stopping or changing mode of an application. The message server distributes its list to all application servers. Before a server is notified of an update, there is a check from the message server list to see if it is available by a dialog server that decides where to send the update given the particular type of notification. Once the dialog server decides where the data is to be written given factors such as workload capacity and code page matching, etc., it updates the decided—upon table located on an available host.

After the dialog server has written the data to the update table, it notifies the update server to process notification. If the update server fails, the message server is notified by an error message, or if a hardware failure a no-response error to the periodic ping by the message server to the update server. After this process, the message server updates its process and availability list and renews it with all the functioning application servers. Then the first available update server responds to the ongoing dialog processes. With this in mind, the system administrator really only has to define the server processes among multiple servers, and the rest of the work is performed by the system due to dynamic update functionality. Dynamic update ensures a higher degree of availability of the SAP system due to the operations of the update servers, and results in a better distribution and balancing of the computing workload.

SAP Elements in Financial Accounting

a. Company—The company is the smallest organizational unit for which there can be accounting operations in SAP, including the preparation of financial statements. One company may have many company codes, each of which is required to have the same chart of accounts as the others.

b. Credit Control Area—the credit control area is the business unit in SAP from which accounts receivable is tabulated. One credit control area may have many company codes and thus many accounts receivable operations linked centrally. In this way it is easy to trace customer credit terms that have been created in a different company code. Operations such as invoicing and dunning in accounts receivable are regulated through the credit control area. The customizing of the relationship between the credit control area and the company codes takes place in the IMG—the company, business area, company codes and the credit control areas are all defined and assigned in customization. Once these elements have been set up, further customization is needed through Financial Accounting Global Settings in the IMG.

c. Chart of Accounts—this directory contains all of the company's general ledger accounts and the chart of accounts is often SAP-delivered. The chart of accounts is used by both the FI and CO modules, so there is one chart of accounts for each controlling area in CO. There can be more than one company code per controlling area, but all company codes in the same controlling area must have the same chart of accounts.

d. Controlling Area—Each controlling area has an operating concern and the controlling area is the basis for management data and reporting in CO.

e. Financial Accounting—The structure of the financial accounting module in SAP contains the following components: 1. General Ledger, 2. Special Purpose Ledger, 3. Accounts Payable, 4. Accounts Receivable, 5. Asset Accounting, 6. Consolidation, 7. Controlling, 8. Investments, 9. Funds Management, and 10. Travel Management. An in-depth analysis of these is beyond the scope of this paper, but for more information on these FI components, see www.saptech.com.

f. Accounts Payable—This component of SAP Financials manages vendor accounting data and payments to vendors. Vendor records consist mainly of the following three types of information: 1. General data, 2. Company Code data, and 3. Purchasing Organization data. The activities in accounts payable relate mostly to the handling of incoming invoices and payments on those invoices; this can be customized to take place automatically through the payment program.

g. Asset Accounting—asset accounting in SAP FI has the following elements: fixed and intangible asset accounting and valuation, leasing, asset consolidation, and a reporting information system. Accounting information is present in SAP Asset Accounting for the asset over its entire life, including self-constructed assets, depreciation, capitalized interest and other related expenses to purchase or construct the asset and put it in place.

Special Purpose Ledger

SAP has some standard ledgers including the general ledger, consolidation, profit center, and reconciliation ledgers. Another type of ledger, the special purpose ledger is a supplement to the standard ledgers that enables you to draw information from different tables for benchmarking purposes. Special purpose ledger can be used if you are not using CO, but want to implement cost center accounting. Customizing a special purpose ledger by taking data from other SAP applications and importing it into SPL must be performed very carefully. The data will have to be validated, etc. and may even be changed during the import process. SPL tables that are set up by SAP can not be used directly as they are only examples of the type of tables you will use in your evaluations. You customize copies of these tables in order to set up your own SPL using dimensions and coding blocks that pull data from cost center accounting, profit center accounting and the general ledger, for instance. The Integration manager is necessary in order to post to SPL's from other SAP or external applications.

Setting up SPL includes defining your posting periods, defining the SPL versions (actual vs. planned, and with or without allocations, for instance), which is followed by defining your validation and substitution rules. Validations are very important in SPL, because almost all data is validated before it is posted to the ledger. Validations are created using Boolean logic and only data that is checked by the validations comes into the ledger. In the IMG, you can customize data correction rules to use in the event there are errors in posting to SPL.

Before you implement SPL, you need to define "sets", or ranges of values in a dimension of a database table, or sums of values in dimensions. The following activities take place in SPL in the order illustrated here: 1. Planning—this is the actual input, distribution and posting of data. The data can be entered directly, or at different levels (such as at the group level, or the account assignment level). 2. Posting—this process takes place in SPL by importing data from other SAP or external applications, or through direct entry for adjustments. The SAP or external data is posted real-time, and data from the external system requires an extractor. 3. Other tasks performed periodically—a. Allocation: allocations and distributions can be treated as fixed amounts, fixed proportions, or based dymanically on set configuration. b. Rollup—using Report Writer and Report painter to read the data and present it can call for rollups to save the data in abbreviated form. c. Currency translation is also possible in SPL. d. Balances from a previous period can be rolled forward into the next period (i.e., asset accounts and certain profit and loss statement accounts). e. SPL also contains some standard data extractors to get data from external applications. f. SPL also has an archiving feature that deletes data from SAP and places it in a file for transfer to storage.

SPL has certain advantages over the standard ledgers, including: 1. Customer—specific requirements for reporting can be customized using SPL, 2. You can determine specifically which transactions are posted to SPL and adjustments can easily be made to the transactions, 3. You can set up separate SPL's for a group chart of accounts or a country—specific chart of accounts, 4. Different fiscal year variants are possible in SPL, 5. Validations and substitutions make manipulation of data easy after the data is transferred from SAP or other external systems, 6. Reporting in SPL uses Report Writer and Report Painter and reporting is possible across applications and organizational units.

Other Features of MySAP Financials

1. Reconciliation ledger—One of the more important areas of FI/CO is the reconciliation ledger which keeps activity in FI in balance with activity in CO. The reconciliation ledger tracks all activity in CO at the summary level (reporting) and activities in CO can be reported through object class or object type. Object types are used in CO to report summary and line item activity: Cost centers, orders, cost objects, networks, reconciliation objects, sales documents, and business processes, etc. Some object types can support many different object classes (overhead costs, investments, production, and profit and sales) and others can support only one object class. Cost centers, and cost objects, reconciliation objects and sales documents have fixed class assignments while the other object types can have classes assigned by end-users: Cost centers will always be associated with overhead costs, cost objects will always be associated with production, a reconciliation object will always be

associated with profit and sales, and sales documents in CO will always be associated with production.

2. Alternative currencies—The currency that is usually allowed in FI transactions is the company code currency only. In customizing the payment program, it is possible to allow for payments to vendors in other than the company code currency. See Financial Accounting > Accounts Receivable and Accounts Payable > Business Transactions > Outgoing Payments > Automatic Outgoing Payments > Payment Method/Bank Selection > Configure Payment Program in the IMG. International companies usually report in the currency of their company code, but you have the ability in FI to translate foreign company financial data if a conversion method has been set up in the foreign company's master data record.

3. Lockbox—SAP Financials support the US lockbox file formats, but make sure which one your bank is using and test its files to find if the format the bank uses is SAP compliant. ABAP programming can be used to conform to the bank's format, and the bank may modify its format for SAP. The formats differ in the level of detail used to process invoice and payment information. One format does not support partial payments or the payment of several invoices with one check. Another format, BAI2, allows as many detail levels as necessary for payments and reason codes. See transaction SE11 for the structure of the BAI2 format. The lockbox can be customized using transaction OB10 where you will need your company code, a name for the lockbox, your house bank information and account number at the house bank. Lockbox processing is set up through transaction OBAY, though customization for the BAI2 format is SAP delivered.

4. Travel Manager—this module, integrated with FI allows you to define legal and corporate per diem rates for travel, meals and entertainment provided to employees by the company. It is possible to define how much an employee will receive by entering different parameters concerning trips, meals and other expenses including accommodations, and vehicle expenses. It is possible also to manage board and lodging costs, meals, day travel, and taxes, etc. for domestic or international business trips. The travel management data is posted directly to FI for payment of trip—cost reimbursements, and corresponding postings are made in CO at the same time. Travel management can be cus-

tomized through Financial Accounting > Travel Management in configuration.

5. Schedule Manager—Part of mySAP Financials that standardizes process management. In this module, templates can be created for various processes in FI, including templates to streamline the closing process. Schedule Manager is designed to minimize errors and reduce time delays by bundling or automating processes in SAP. It also enables automated collaboration in workflow and interactive features to reduce and correct errors.

A Brief View of Archiving

Data archiving is the process of transferring data from your real-time database to a different form of storage such as disk or tape while maintaining all or part access to the data as required by your business process. Document archiving may consist of scanning documents and storing them, or archiving outgoing documents in a PDF format, for instance. Data archiving has significant benefits, including reducing storage costs, enhancing system availability, performance and resources, systematized records retention, and a reduction of downtime during upgrades. Data archiving may also identify new business processes or ways to improve them.

When some IT people think about archiving, they think mostly just about adding hard disks to the database environment. The growth of the database under this scenario, even with data archiving is significant and can adversely affect system performance and business processes, etc. What to ask yourself is—a. Is the data really still needed or can it be summarized, and b. Can the data be archived in order that it not stay in the real-time database? This is called data prevention. Data prevention seeks to minimize unwanted growth of the database.

In order to archive data in SAP, you need the Archive Development Kit that handles the adjustment of code pages, number formats, data structure changes, data compression, file handling and job scheduling. For the ADK, see transaction SARA that will take you through the process of preparing for the archive, the archiving itself, deletions and post-processing, and reporting, etc. Archiving consists of compressing business objects from the following tables (and more): BKPF, BSEG, BSIK, BSAS, BSAK, and so on (for tables and indices, see transaction DB02; for tables only, see DB15). The business objects have archive objects with a special nomenclature that identifies them as FI, MM, SD documents, for instance. Archiving objects have retention times and storage parameters, as well as other features that are customer-defined. Archived data can be stored in several forms such as file or tape form, or in a form specified by third-party software. Archived data can be accessed from optical disk, tape, file or from a third party through SAP transaction ALO1.

In order to have the archiving project run smoothly there must be a commitment from management to the process, the archiving must be handled by an experienced project group and technically oriented project manager, archiving must have priority (sometimes) over more glitzy projects like CRM or Web Portals, and necessary steps must be made to properly handle data that must be accessible to end-users after the archive. These points can usually be attained through use of internal e-mails, good recruiting, strong leadership, and an engaged project team.

Archiving is more difficult if you are working with older versions of SAP, and in this case you should only use the most common SAP objects including IDOCs, Financial and MM documents and Workitems. Your Basis team should have expertise in this area and needs to work with the functional team in each module.

In order to assure a successful archiving process, you must have, among other things: a. Transport documentation, b. an OSS log, c. White papers pertaining to your processes, d. Training material for the project team, and e. Documentation of object details and customizing details. Testing should also be performed repeatedly and should be documented. Customizing should also be tested in SAP and the archive system. For further information, consult the SAP Library, SAP Online Help, OSS notes for archive objects, and read Archiving Your SAP Data by Helmut Stefani available through SAP Press. There is also an Archiving Discussion Forum at www.ASUG.com and more documentation available through http://service.sap.com/data-archiving.

A Brief View of Authorizations

The R/3 system has authorizations in order to define and assign different user functionalities. Authorizations in SAP include the following capabilities: 1.

An end-user Profile Generator, 2. The feature of locked or unlocked transactions and records, 3. Data encryption, 4. Locking for changes and 5. Structural Authorizations. SAP authorizations enable the assignment of general or finely detailed roles or duties for each user. The authorizations are centrally maintained in user master records and enable the handling of SAP modules in the areas of specific operations of the enterprise. Each user may have several or many, many authorizations and the relationships can become complex, if not problematical and conflicting in some areas. Authorization objects, therefore, must be a subject of painstaking care.

Each authorization object is a combination of different fields that may or may not have intervals attached to them. Authorization checks protect the function of objects in the system and SAP has authorization logic built in to its applications. Authorization administrators need to decide the functionality of the checks and how the check should be conducted in addition to the exact parameters for user profiles. The Profile Generator can be used to generate authorizations and user profiles, and authorizations can be manually entered into a user profile. Developers are primarily responsible for creating new authorization objects in a client—independent environment through the ABAP Workbench (ABAP Workbench > Development > Other Tools > Authorization Objects > Objects.)

You can set up the Profile Generator using transaction SU25, but first make sure the R/3 system parameter is correctly set to active. Users are created using transaction SU01. You can also use this transaction to create a template user, copying users, and entering passwords. Create user groups using SUGR (Tools > Administration > Maintenance > Maintain User Groups.) To find which users are authorized for a specific transaction, go to Tools > Administration > User Maintenance > Information System > Transaction Authorizations and enter FB01. SAP also delivers many, many user role templates: To work with User Roles, go to PFCG and follow the menu path Tools > Administration > User Maintenance > PFCG—Activity Groups. Then select an activity group, and select "Display" to see the authorizations for this group. With guidance from an authorizations person, you can also "Change" or "Create" authorization groups in this series of screens. Remember that you can assign user roles through either transactions PFCG or SU01. You can also assign user roles by clicking on "Other Menu" in the "User Maintenance" screen.

Tax Reporting and Compliance

Customization for sales and use tax in FI is made in the IMG at the country level. What follows is a brief illustration of U.S. sales and use tax customization through IMG transaction OBBG (Financial Accounting > Financial Accounting Global Settings > Tax on Sales/Purchases > Assign Country to Calculation Procedure.) There are two ways to configure your sales and use tax: 1. Using procedure TAXUSJ where the tax parameters are set up and maintained manually, and 2. TAXUSX where third party software is used to set up sales and use tax. In setting up the structure for your tax jurisdiction codes, you have up to four levels of customization including state, county, city and other (for special circumstances). In order to define and assign your jurisdiction, go to transaction OBCO (Financial Accounting > Financial Accounting Global Settings > Tax on Sales/Purchases > Basic Settings > Specify Structure for Tax Jurisdiction Code) and do the following: a. Enter the name of the tax procedure for your jurisdiction, and b. Enter a description for the jurisdiction. You must then enter the character lengths of the first through fourth levels in the hierarchy of your tax jurisdiction. Check the "TX" box if you want to determine tax on a line-by-line basis instead of on a cumulative basis.

SAP also has functionality for VAT, and input and output taxes under "Basic Settings" in the Tax on Sales and Purchases area of the IMG. This area customizes calculation routines for each jurisdiction, you can also add condition types to include several types of tax in this area. In addition, withholding taxes can be customized in the IMG under Financial Accounting Global Settings > Company Code > Withholding Tax. Withholding taxes can be either "simple" or "extended." Simple withholding taxes are items like creditor withholding, and withholding for payment. Extended withholding taxes can be customized when there are several withholding taxes per creditor, withholding for partial payments, or calculation of the tax when posting an invoice and on payment of the invoice.

SAP FI Reporting

This section includes a brief taxonomy of some of the thousands of SAP-delivered standardized reports. This list does not contain all of the FI reports, but does touch upon the more important ones.

a. General ledger account listing—SAP menu > Information Systems > Accounting > Financial Accounting > General Ledger > Information System > Master Data > G/L Accounts List > G/L Accounts List.

b. General ledger line items—Transaction F.51, then fill in your parameters.

c. General ledger account balances—Transaction F.08, then fill in fields for your query.

d. General ledger balance sheet and profit and loss statements—Enter transaction F.01. For this transaction to run, the proper financial statement version must already have been created.

e. Customer open items report—At the command line, enter transaction S_ALR_87012178.

f. List of customer open items—At the command line, enter transaction S_ALR_87012173, choose execute and fill in the fields for the desired query.

g. List of vendor open items—At the SAP menu, go to Information Systems > Accounting > Financial Accounting > Accounts Payable > Open Items. Make sure a date is selected for the open items key date (system date is the default) before running this report.

Strategic Enterprise Management and SAP FI

In the past competition between companies was focused on products and differentiated product markets that were limited in scale and scope. This called for an emphasis on internal processes that had to do with development, manufacturing and selling on a local level. The operational success of the local operations of any enterprise was seen as crucial to competitive advantage and success. The ordinary measure for economic success of the enterprise was the profits generated by the selling of its products. This paradigm, the inside-out approach, that concentrated on the historic success of the enterprise regardless of external influences, and a financial approach to short-term tactical management with little understanding of eventualities, has become outmoded.

Competition in the product and service markets has intensified greatly as economics is more and more globally oriented. Product life cycles have shortened significantly creating a demand for better research and development processes. Circumstances in the capital and labor markets have also changed, not in just products and services. The business enterprise today must be more responsible to its stakeholders through adding value to its core processes and increasing total return to shareholders. Because there has become increased linkage between strategy and execution, a vision for the organization, and value-based management, the major focus of the SEM environment are these factors, and other drivers of shareholder value.

By analyzing the likely cash flow of an enterprise, SEM can determine which strategies are likely to generate an acceptable rate of return and future value potential for the business as a whole, or for individual business areas. In SEM, this process is integrated in strategic, financial and operating information in the following areas: a. Business Information Collection—this function automatically sources unstructured business information from the world wide web; b. Business Planning and Simulation links strategic planning and simulation with cross-functional (application) enterprise planning; c. Business Consolidation can be used for financial and management reporting consolidation; d. Corporate Performance Monitor is a system of performance and strategy communication and monitoring through Balanced Scorecard, Management Cockpit, and Value Driver Trees; e. Stakeholder Relationship Management integrates important influences and individuals into the enterprise management process. The SAP SEM application is underpinned by the SAP Business Information Warehouse, and all functions of the SEM components can operate on one or several SAP BW infocubes.

Integration of a strategic process could have the following characteristics on the corporate level: 1. Enterprise Modeling calls for a definition of the enterprise structure through activity models and key performance indicators. 2. A collection of internal and external information, coded or not, through Business Information Collection would add to the development of a strategy. 3. Business Planning and Simulation for strategy and enterprise planning might call for market modeling and simulation, business strategy modeling and simulation, enterprise planning, and risk—adjusted value calculation. 4. Business Consolidation would control legal and management consolidations, valuation adjustments and economic profit considerations. 5. Corporate Performance Monitor would analyze performance, monitor key performance indicators and the Balanced Scorecard, and management would confer over information supplied in the Management Cockpit. 6. Strategic actions, when taken, could be monitored and tactical targets could be changed, and enterprise structure could be changed and communicated to stakeholders—this would all be facilitated by the Stakeholder Relationship Management function. These activities could be changed or monitored down to the desired level of detail, and all the information processing would be supported by Business Information Warehouse (BW).

The final product of SEM is communication of strategic vision and information to stakeholders in order to accurately disclose the value of the enterprise (assets and intangibles) and its competitive position. Shareholder value can be enhanced by using the SEM processes and increasing communication with and knowledge about the external market. With this in mind, inconsistencies in economic value will be reduced, corporate governance will improve, and communication with shareholders will increase public confidence in the corporation. SEM in its integrated information system presents a series of processes to stay on the road to consistently adding value and increasing the economic reach, efficiency and effectiveness of the enterprise.

A View of R/3 Enterprise Financials

Business establishments today are striving for more efficiency, performance and control using integrated applications that lead to increased profit growth over time. SAP R/3 enterprise has the integration features, ease of use and upgrades, ease of operations and customer support to enable the management of systems and people at almost all economic levels. SAP R/3 enables a prime IT competitive advantage as it becomes and remains a successfully implemented and utilized enterprise system.

SAP R/3 system upgrades will be more flexible going forward, and a complete set of business applications will be offered with an integration platform that will make for a more adaptive business model. The mySAP Business Suite will be based upon SAP Netweaver and extensions in an application infrastructure. R/3 Enterprise will be based primarily on a web application server, and SAP Financials are part of the R/3 Enterprise Core supporting Business Suite, SEM, and even portal solutions. It is estimated that SAP Consulting will take approximately six months to upgrade a system to R/3 Enterprise.

Among the enhancements to SAP Financials in R/3 Enterprise are the following: a. All current functions of FI are extended into R/3 Enterprise Financials with the possible exception of asset accounting; and b. Enterprise Financials have an Accrual Engine that will enhance accounting utility for contract business transactions, transactions like bonds and subscriptions, and any periodic postings based on amount or value. The Accrual Engine is a kind of accounting subledger that supports multiple GAAP's, has a simulation tool, and has reporting tools that integrate with BW. Asset accounting will change under Enterprise Financials as account assignments are possible for additional objects in real estate and public sector, etc. Additional account assignments for Asset Accounting are available in the Asset Accounting section of the IMG (Asset Accounting > Integration with the General Ledger > Additional Account Assignment Objects). The asset depreciation program (RAPOST2000) has also been redesigned as additional account assignments are available, batch processing is no longer necessary, and depreciation accounting is integrated with Schedule Manager. Depreciation postings are also possible for more than one company code through transaction RAPOST2010.

Asset Explorer has been improved: a. With enhanced searching properties (enter an asterisk in the field sub—number and you can view the main asset and sub numbers); b. With transaction simulation; c. With a simulation for a change in depreciation terms; d. You can also refresh displayed data without re-starting the transaction in asset explorer.

The Audit Information System now facilitates smoother and better quality audits, and consists of collection, structure and default setup roles of standard SAP programs. This auditor's toolbox now has a reporting tree that mirrors the balance sheet, and evaluations with control variants are available for every auditing function.

The automatic payment program in Enterprise Financials will enable automatic analysis of the payment proposal and automatic block or automatic settlement of payments to vendors. It replaces external software use and is integrated with Schedule Manager.

R/3 Enterprise also has a Document Relationship Browser that reveals the structural and management hierarchy of documents in the system. SAP 4.7 also supports Profit Center documentation, and has enhanced exchange rate customization available in the Currencies section of the IMG. R/3 Enterprise also has enhanced the 4.6C Data Medium Exchange so that no ABAP knowledge is necessary for end-users. The DME Engine also integrates into the Payment Medium Workbench and notes to payees are customized rather than hard coded.

SAP 4.7 provides new functionality with leading—edge technology. It is a flexible system that will be developed further into the future. The system, based upon a web application server has a core and extensions with improved stability and upgrades on different applications can take place as you need them.

Tips, Commands and Menu Paths

1. Account Groups—These simplify the creation of G/L accounts and are defined per the SAP-delivered charts of accounts. They are mandatory for G/L accounting and allow for an internal or external number range for A/P and A/R, etc. Account groups are defined through the following menu path in the IMG—Financial Accounting > G/L Accounting > General Ledger Accounts > Master Records > Preparations > Define Account Group. Also see transaction OBD4. For Accounts Receivable and Accounts Payable, if you want to change an account group, you need to create a new master record.

2. Account Types—There are five that are allowable: D—debit, K—credit, S—everything else, A—asset, and M—material. These must also have number ranges and account types and different from account groups which are master data. See the IMG in Financial Accounting > Global Settings > Document Header > Overview. The posting key controls exactly which account type is used. The posting key look to see whether

the entry is a debit or credit, and posting keys control the fields you see (field status groups). Never create new posting keys in the IMG and pay attention to the field status variant in Maintain Table Views. Posting keys represent horizontal control of what you can and cannot do and see. You can have a posting key driving what is required or not in a G/L posting, the same for field status. You can also have "Activity Based" controls such as create, change, or delete.

3. Accounting View—Transaction AC19.

4. Accounts Payable Master Data—when you create a vendor/customer, you must put them in an account group with a company code. Externally numbered vendors are alphanumeric and internally numbered vendors are numeric only. Menu path for vendor master data: Accounting > Financial Accounting > Accounts Payable (Receivable) > Master Records > Create.

5. Advanced Planning Optimizer—a new dimension "bolt-on" product that enables you to look at supply requirements for all plants together.

6. ALV—ABAP List Viewer.

7. Application Link Enabling—for distributed systems, and like EDI but better—ALE enables you to move master data, etc. across different systems.

8. Asset Class—essentially an account group in FI that categorizes assets and influences screen display, and account determination works out the postings for bookkeeping procedures. Using asset classes, you can track multiple valuations for the same asset. Assets are put in a class and then are grouped into a Group Asset. Menu path for creating an asset: Accounting > Financial Accounting > Fixed Assets > Asset > Create. Be sure to include the asset's account group, business area and cost center; save, and SAP gives you an asset number.

9. Assignment—Transactions AO90 and AO86.

10. Authorizations—Roles are specific settings for users and fall under the rubric of "Purchaser", "Preferred Purchaser", "Controller", "Assistant Controller", etc.

11. Automatic Payment Program—The following is necessary in order to implement automatic payments: 1. Define your parameters (configure the payment program in the IMG; you also input some parameters at run time), 2. Create a payment proposal (select open items for payment), 3. Payment run—creates an FI posting, 4. Print the payment medium (check, EDI, floppy disk, etc.). Transaction FBZP configures the payment program.

12. Balance Confirmation—for accounts receivable, go to transaction F-17.

13. Banking with Accounts Payable—You must have a bank master record that can be imported into the bank directory and into the bank table, and for automated transactions with vendors and customers, you have to update "Client" or general views. Menu path for finding vendor bank information: Accounting > Financial Accounting > Accounts Payable (Receivable) > Display. There is an IT shell for each bank. In the IMG: Financial Accounting > Bank Account > Bank Account > Define. You must have the bank key (routing number) in the IMG Define screens. Once the bank account is created, you can create other accounts within the same bank using identifiers. As you create the separate bank account, give it an account identifier including the bank account number, currency and the G/L account it will post to. Your master data records allow you to look at bank details from the accounts you have created.

14. Batch Input—From the menu bar, choose System > Services > Batch Input; highlight the batch and click "Process". A pop-up box will tell you when the process is complete. To find the last document posted: Menu path—Financial Accounting > General Ledger > Document > Display.

15. Binoculars button on the main menu bar is used to "Find" menu topics and commands.

16. Business Application Programming Interface—Only one BAPI per business object.

17. Business Area—Maintain a business area using transaction OX03. One business area can have many companies. In Controlling—are entities that run across divisions and are defined in FI—a business area is assigned to a transaction by way of rules in the Logistics Tables (SPRO).

18. Business Warehouse—acts as a huge central depository for information and contains standard extractors—not real time. BW plug-ins work with R/2, a flat file, Peoplesoft, etc., and BW summarizes transactional data.

19. Cash Discount—Define a cash discount for accounts receivable in transaction OBXI. In accounts receivable it also may be necessary to define reason codes for different types of under/overpayments using OBBE and OBXL. Days in arrears can be automatically calculated, provided you maintain this program through OB39.

20. Cash Position Report—Transaction FF7A. Transaction FF7B gives a liquidity forecast.

21. Catalogs and internal in R/3, or external on the web for which you will need an ITS (internet transaction server for version 4.6C)).

22. Central View of Master Data—a combination of the general view, company code view and MM view (Materials Management). The Sales view is a combination of the general view and MM view. The message "Material does not exist" may mean the **View** does not exist.

23. Chart of Accounts—Create a chart of accounts with reference to another company code using OBY2.

24. Check Register—Menu path: Financial Accounting > Accounts Payable > Environment > Check Information > Display. Transaction Code FCHN. Define the number ranges for checks using transaction FCHI. For voiding checks, define the void reason codes at transaction FCHV.

25. Clock button on the menu bar is the executable icon.

26. Closing periods in FI—there are sixteen closing periods in FI: 12 plus four extra in the twelfth period. FI needs a fiscal year or calendar year variant for closing purposes. For non-calendar years, create a variant with non-calendar periods and assign it to a company code. The fiscal-year variant is assigned and maintained in the IMG Global Settings under Financial Accounting > Global Settings > Fiscal Year; select your option, then "Periods". The fiscal year variant screen is found at OB29.

27. CO—You can take all of your company codes and fit them into one CO area.

28. Commands and Transaction codes—how do you find them: From the main menu—System > Status.

29. Company—To create a company, use the IMG: Enterprise Structure > Definition > Financial Accounting > Define Company Code.

30. Company Code—Represents a legal entity. Where possible, copy the settings for a company code from a similar or identical company, the same for plant set—up. The company code creation screen is at OX02. Copy a company code using transaction EC01. Find the Organizational Object Company Code blank screen, select the copy button and wait, input the company code to copy from and follow the screens. Expect hourglass time. The idea is to create one company in the IMG through the request screen and do not overwrite it or modify it. Create other companies from it as they will automatically have its attributes. After you copy the company, check that its company code exists in the IMG. In the IMG, use "Change Country Specific Settings" to modify attributes for a specific country.

31. Consolidations—Link the lower-level accounts to the consolidated chart of accounts (CONS) by mapping them through FI.

32. Controlling Area—CO space for which you can have more than one company code. Same chart of accounts as the company code and same year-end (fiscal year). You can have more than one controlling area per client. If you use COPA in your controlling area you need an operating concern.

33. Country Chart of Accounts—Assign a country chart of accounts through OBY6. Copy chart of accounts through OBY7. Display the chart of accounts using transaction FSP3.

34. Credit Management—The credit control area data is in OBY6 (master records, credit limits, credit management). Credit management is an integration point with SD and can be controlled based upon a total amount of credit, and individual credit limit, etc. The company codes must be

mapped to the credit area (OBY6). Create a credit control area using transaction OB45. Assess credit risk categories by following the menu path to Financial Accounting > Accounts Receivable and Accounts Payable > Credit Management > Credit Control Account > Define Risk Categories. You can also create employee credit representatives and group them accordingly.

35. Customers and Vendors—Make sure the customers and vendors are in the same account group (D for Accounts Receivable, and K for Accounts Payable)—this should be embedded in the master records. Vendor account groups can be created using transaction OBD3. Send vendor master records to another company code using FK15. You can also group customers (transaction OBD2) and vendors.

36. Depreciation Posting Rules and Intervals—Transaction OAYR. There are three types of depreciation in SAP: Ordinary, Special, and Unplanned.

37. Document Type—Document type ranges can not overlap. The FI document type controls the type of document in CO. There are no document numbers in CO. Any document must have a "document type" in its header (define document types using transaction OBA7). Documents must have numbers and ranges attached to document types (define the number ranges using FBN1); a range of allowable account types in documents must also be determined. The standard document types are DR, DG, DZ, SA, KR, KG, KZ, KN, AB and G/L documents are generally of the SA type and must have a number range assigned to them. In the IMG, make sure you attach a number range key to a document type.

38. Downpayments—Transaction F-37 (downpayment request), F-29 (actual downpayment), followed by F-39 (clearing). The downpayment will clear the downpayment request on the customer account. When you go to F-39, you don't have to clear the entire downpayment as partial clearing is available. Downpayment processing in accounts payable—F-47 (request), followed by F-48 (to record the downpayment), and then F-54 (to clear the payment).

39. Edit a General Ledger Account—Transaction FSOO.

40. Encashment Process—Transaction code FCHR.

41. External Reporting Charts of Accounts—Chart of accounts CAUS and chart of accounts CAFR are mapped together for external reporting. For internal reporting purposes CAUS is mapped to group chart of accounts INT. If you change the chart of accounts configuration when you are configuring the company code, all company codes that use that chart of accounts will also be changed. After you create your chart of accounts, you must transport it across all company codes using OBY9: Financial Accounting > General Ledger Accounting > Data Transfer > Transport Chart of Accounts.

42. Extras > Settings—Adjust you own user settings.

43. FI Configuration—Transaction ORFB.

44. FI Global Settings—Menu path: Financial Accounting > FI Global Settings > Company Code > Enter Global Parameters: This is the place to configure for Business Area Financial Statements, turn on cash management, set fiscal year-end and assign the chart of accounts (fiscal year variant).

45. Field Status—1. Create, 2. Change, 3. Display. Field status is determined in the IMG. Define field status groups using OBC4.

46. Field Status Definition—S—suppressed, R—required, and O—optional.

47. Field Status Groups—These represent vertical control linked to accounts (when you display a G/L account centrally) and are linked to a field status variant. Hierarchy for the field status groups is as follows: a. Definition (S, D, R, O)—don't use the combination of S and R in a definition, b. Group, c. Variant, d. Company Code (OBY6). In the IMG, see Financial Accounting > Global Settings > Document > Line Item > Controls > Maintain Field Status Variants. Field status groups are found in the Master Data in the general ledger.

48. Financial Statement Items—Transaction FSE2.

49. Financial Statement Versions—Transaction OB58. Financial Statement versions group together certain accounts in FI for financial statement pre-

sentation purposes. Some financial statement versions are SAP delivered, but you can define your own by going to Financial Accounting > General Ledger Accounting > Business Transactions > Closing > Document > Define Financial Statement Versions.

50. Folder button on the menu bar will launch documentation.

51. Functional Areas—Functional areas in SAP are spaces such as administration, marketing, production, and sales, etc. To use functional areas, items necessary in SAP, use transaction OKBD. Substitutions are used in functional areas and they populate table fields with necessary data behind the scenes in the posting process. To define substitutions, use transaction GGB1. To simulate a substitution once you have created it, use transaction GGB1.

52. General Ledger Accounts—Create a general ledger account using FS01.

53. General Ledger Posting—Transaction FB50. Here you can transfer, and use different company codes and this transaction has company code and central views. Intercompany postings create three documents: a. A posting from the first company code, b. A posting from the second company code, and c. A clearing account document. Automatic account assignment in intercompany postings is defined in OBYA.

54. General Posting Complex Screen—FO2.

55. Global Parameters—If you are looking for a configuration item and you have no idea where it is, look in Company Code > Global Parameters.

56. Goods Receipt—Transaction MB01.

57. Group Currency—Set this using transaction SCC4, and OY01 for additional currencies. Delete local currencies in the IMG with OBS2. The object currency is the default company code currency.

58. House Bank—Defining a house bank is like setting up master data using bank—provided information (be sure to have a routing number and a SWIFT code as well as a G/L account)—use transaction FI12.

59. IMG—Using the IMG, you can break different activities into different projects.

60. Intercompany Journal Entry—this is possible provided both companies have the same chart of accounts and the same fiscal year variant. The internal clearing accounts are set up in the same area of Configuration as the discount accounts (or transaction FBKP, or System > Services > Table Maintenance and go to Table 30). To set up intercompany clearing accounts go to FBKP, click on automatic postings, and go to "Company Code Clearings" to access clearing between company codes, then fill in the dialog box.

61. Manual Accrual—Transaction FB51. Remember to input a reason code, otherwise it's just a journal entry. For a reversal, use F.81.

62. Master Data—FI View: Company Code and Vendor or Customer.

63. Master Data in Travel Management—Employees are set up as vendors for reimbursement purposes—integration point with HR. You have to have mini—master records for Travel Management to work.

64. Material Document—Material Receipt documents create also documents in the following areas: FI, CO, MM, SPL, and Consolidation, and possibly other areas such as Profit Center Accounting depending on configuration.

65. Navigate—\n after your transaction entry lets you go from screen to screen.

66. Open Items and Posting with clearing—Use F-04 through the general ledger, F-30 when you are going through accounts receivable, and F-51 when going through accounts payable. All three of these transactions have the same screens. Transaction FB00 is available in order to have edit options for clearing and other functions.

67. Original Document—Environment > Original Document carries you back to the original document in the FI view.

68. Parked Documents—See report RFSOP000.

69. Payment Program—The payment program in FI accounts payable allows you to define all parameters for paying vendors. This is where you tell SAP how to process payments and is found in transaction FBZP. Payments in SAP may be processed through a separate company code; therefore, company code data will have to be entered in this transaction.

70. Payment Terms—Payment terms can be created in the IMG to your heart's content: Financial Accounting > Accounts Receivable and Accounts Payable > Business Transactions > Incoming Invoices > Credit Memos > Maintain Terms of Payment or transaction OBB8. See transaction ZBO1 under Change View Terms of Payment Details. Once you change it here, your changes go to the Master Record. There are three ways to apply payment terms in credit memos—reference number, with a blank (not valid), with a "V" which means the payment terms are valid and should apply.

71. PFCJ—Acronym for Profile Generator that generates Roles; each of the steps in PFCJ that makes up a role is considered a task.

72. Plant and company code—One plant per company code. One to many storage locations per plant.

73. Posting Control—some defaults can be set through OBY6 and others under SPRO (Financial Accounting > Global Settings > Documents > Default Values for Document Processing). See the Default Values posting key under Default Values Processing. In order to change text, payment terms etc., in the IMG, see Financial Accounting > Global Settings > Document > Line Item > Document Change Rules (find the item and double-click it).

74. Posting Keys—posting keys in SAP FI determine if the item posted is a debit or credit and what its field status is. Posting keys are SAP delivered and it is not recommended that you modify them.

75. Posting Period—See the company code lookup in order to find out what account type, variant and what it is doing. In the IMG, see Financial Accounting > Global Settings > Posting Periods > Open and Close Posting Periods. There are 16 posting periods in all and posting authoriza-

tions must be assigned. The posting period variant configuration screen is found through OBBO.

76. Posting Using Reference—Look up the document, copy it into the work area, make changes, and post.

77. Purchase Orders in R/3—Utilization of "Enterprise Buyer" and the web to use and set up workflow: 1. PO created and sent, 2. Goods receipt, 3. followed by any other document related to the item including FI and CO documents. With Enterprise Buyer any procedure generating documents is performed in R/3.

78. Purchasing Organization—one for each plant and if you want to use MM you need a purchasing organization.

79. Purchasing View Setup—In order to generate PO's, you must have MM. A PO is not an accounting document in SAP.

80. Reconciliation Account—In SAP you can not post to reconciliation or Control Accounts. Posting is based on the Master Record assignment and customers and vendors are rolled up into the reconciliation accounts. Display the account centrally and click on the Control Data Tab; if the reconciliation account for the account type field is populated indicating what the control account is for, then the account is a control account and can not be posted to. The control accounts can be used in customer or vendor master data for reporting purposes. Posting to the General Ledger is controlled by the reconciliation account.

81. Recurring Entries—a. Build your template, b. input debits and credits and amounts, c. enter the range of dates for which the entry is good and when to launch. Once you have your template, you get a document number that sets up a payment to a vendor as a payable. If the recurring entry is set up properly and you try to execute a payment where it has previously been run, you will get an error message. See Accounts Payable > Document Entry > Reference Document > Recurring Document. To display the recurring documents, use the previous menu path and add > Display.

82. Report Painter—To create a report, use transaction GRR1.

83. Reporting—In R/3 every single application has an information reporting system with standard reports in it (push systems—real time). Business Warehouse is a pull system for reporting and reports are run once a week maybe. Many SAP users use the reporting systems in R/3 to monitor transactions and use BW for reporting.

84. Retained Earnings—SAP gives you the option to have multiple retained earnings accounts, though each G/L account can only have one retained earnings account. Define the retained earnings account using transaction OB53.

85. Reversing Entries—You can use a negative debit instead of the traditional way. This has to be allowed by OBY6 and the document type. There also has to be a reason code (set up in the IMG) for the transaction.

86. Rules—establish a name and a key under Enterprise Structure Configuration in the IMG—define and assign.

87. Sales Organization—One sales organization per company and this can be joined to a distribution channel to form a distribution chain in SD.

88. SAP Client—has its own tables, own master data, and represents the highest level of the SAP hierarchy.

89. SAP Workplace has mini applications that you can build your own windows with, it also works with SSO (single sign—on if you wish).

90. System Help—this button will always appear in the menu bar where the other buttons will change.

91. Table Maintenance—System > Services > Table—Sometimes in an error message the message will tell you which table to maintain (use the table number in the command line onscreen to find the table).

92. Tax Jurisdiction—To specify the structure for your tax jurisdictions, use transaction OBCD. To define a tax jurisdiction use transaction OBCP. Taxes on sales and purchases are defined in transaction FTXP. Non-taxable transactions are defined in OBCL. Automatic account assignment for sales and use tax transactions is defined in OB40

93. Tolerances—In the IMG, see Financial Accounting > Accounts Receivable and Accounts Payable > Business Transactions > Incoming Payments > Global Settings > Define Tolerance Groups for Employees (Vendors, Customers). For manual payment receipt configuration, use transaction OBA3.

94. Transaction Code for IMG—SPRO; menu path—Tools > Accelerated SAP > Customizing > Edit Project.

95. Transaction codes—SE93 gives a table listing transaction codes—FI transactions start with F, and CO transaction codes start with K.

96. Two major concepts that SAP revolves around—1. Views, and 2. Groups. As configured in the IMG, groups are customers, accounts, etc. The idea is to group everything and build a hierarchy of groups. Many of the objects are matrices and a view is a cell in each matrix (the object exists, the view does not).

97. User settings—System > User Profile > Own Data—see your settings.

98. Validations—Validations are used to check and confirm configuration settings according to your needs in FI. See transaction GGB0 for validation configuration, and activate your validations using transaction GGB4. To simulate a validation once you have created it, use GGB0.

99. Variants—Goto > Variants > Save as Variant—to make or save a variant; or Goto > Variants > Get—to retrieve a variant. Variants can run your report or transaction in the background, change formatting from the standard, can protect the report or transactions and other attributes. Variants can be assigned to objects such as a company code or posting period.

100. View—There are three views—a. General data view, b. Company code view, and c. Functional view (purchasing, sales, FI, etc., and there are different levels obtained within the module view).

101. Voiding Checks—Transactions FCH3 and FCH9. Transaction FCH8 voids the check and reverses the payment document.

Chapter Two

Financial Supply Chain Management and mySAP Accounting Tools

Financial Supply Chain Management

Financial supply chain management for SAP begins with a set of tools to better coordinate the cash cycle of the enterprise. These tools include a direct billing and consolidating function, a credit control function, and bank and cash management tools as well. Financial supply chain management relates to all transactions and activities in the cash flow process starting with the customer order through account reconciliation and payment to the enterprise. The idea of FSCM is the augmenting of a value—chain between businesses and in intracompany functions and processes that begin before buyers and suppliers meet and that continue after the time of settlement. FSCM also facilitates the flow of financial information between businesses and within an enterprise.

In the shortened cash cycle times of the modern marketplace, FSCM enables economic exchange through financial support, information dissemination and analysis in the areas of product quality assurance, corporate finance, and marketing that lead up to the fulfillment of product orders. FSCM facilitates the settlement process through accounting and credit management tools from invoicing, to review, disputes and payment. It also allows for an efficient and accurate analysis of the credit and cash management areas of the business. FSCM can solve financial difficulties such as: a) Large amount of uncollectible accounts, b) High financing costs with external banks, c) High costs of processing paper and high maintenance EDI systems, d) Lack of financial integration with business partners, e) High percentage of disputed accounts, slow payment, and poor customer relations, f) Expensive payment processing services, and g) inaccurate cash flow forecasting, tied up cash and poor cash management in general.

FSCM has a systemic flow from supplier to buyer called order to cash, and a flow from buyer to supplier called purchase to pay. Order to cash includes accounting, cash and accounts receivable management and logistics features. Purchase to pay includes accounts payable and inventory management as well as cash management features. Optimizing these flows is important because it can reduce float, improve billing and payment processes of a business, and it enables easier cash management and more accurate cash forecasts. Ineffective cash management is costly and a well implemented FSCM system can increase the operating margins of the enterprise, optimize the collections process and reduce accounting and operational errors.

FSCM actually starts with the billing process for which SAP has a solution for electronic bill presentation and payment called Biller Direct. Biller Direct eliminates many of the procedural steps in the process of generating invoices in accounts and turning them into cash. Biller direct does away with the paper trail involved in printing bills and reminders and sending them to customers, searching for bills when payment comes in, posting problems, and data entry problems. The biller service provider under order to cash allows for an information system to monitor billing status so the biller can review accounts receivable and receive information about billing status for customers.

In Biller Direct there are fewer steps to the collection process and the payment collection process is more rapid, accurate and results in higher levels of customer satisfaction. Efficiencies for the supplier include: a) reduced bill reviewing time before payment, b) increased customer loyalty, c) reduced billing and payment transactions costs, d) Improved payment processing, e) a reduction in outstanding invoices, f) data integration that leads to better business integration, and g) many fewer problems as presented by EDI. Buyer efficiencies created through Biller Direct include: a) A streamlined invoicing and collections process, b) better control of payments and receipts, c) economic profit from better cash management, d) accounts payable data integration, e) reduced accounting and EDI errors, and f) worldwide access to accounts at anytime. Overall, Biller Direct results in substantially reduced billing costs, improved cash flow, facility with electronic trading, better business practices and competitive advantage through better cash management, reduced complaints to the call center and individualized marketing capabilities. Biller Direct is optimal for an enterprise that has decided to deal with electronic bill presentment and payment, wants a simple implementation, and wants to advertise online or conduct a marketing strategy online. Biller Direct requires at least one SAP instance and R/3 4.6C.

SAP also has a tool for the billing and collection process called the Biller Consolidator. The Biller Consolidator creates an electronic network between biller and customer banks with a clearing feature, and between biller and customer entities. The Biller consolidator has a biller service provider that provides an electronic bill presentment and payment service. Under a purchase-to-pay scenario, SAP has a customer service provider information system for enrolling customers, bill presentment and customer service. The customer service provider in SAP allows customer support for billing, payment orders and other notifications including a limited account management tool available to customers. The business process in the Biller Consolidator is as follows—a. An electronic document is sent, b. an invoice is presented, c. payment is released, d. status notification, e. released payment notification, f. payment order is sent, g. confirmation of payment, h. credit advice received, and i. automated account update process. These steps, while performed electronically, reduce billing and transaction costs, increase internal controls, enhance the ability of management of accounts, eliminate possible data entry errors, streamline the workflow process and reduce accounting errors. The system can also be accessed twenty four hours per day. This system is available for the enterprise that has multiple SAP releases or mixture of SAP and legacy systems, that needs to consolidate information from multiple systems, and that is interested in centralized payment online.

SAP FSCM also has a tool to manage bill disputes and to streamline the collection or resolution process called Dispute Management. Dispute management can solve reconciliation and collection problems as well as reduce transaction and processing costs and decrease the amount of outstanding accounts. The process involves the following—a. open the dispute case, b. import the data including master data, c. sending of workflow messages to the responsible party, d. creation of a notes history and a dispute clarification case, e. checks by appropriate personnel, f. dispute status changes, g. dispute case updated and closed.

R/3 can also automate credit decisions to improve efficiency and flexibility in your credit operations with Credit Management. SAP Credit Management is an information system connecting SD, CRM, legacy systems, FI, external credit management tools, BW and the other applications in SAP FSCM. Another feature of R/3 FSCM is the enabling of customers to settle payments internally among subsidiaries through Inhouse Cash. Inhouse Cash is an internal electronic payment exchange network that allows physical payments across international borders. Inhouse Cash is centrally located, usually at the parent company headquarters and where the accounting system is centralized. It allows for—a. electronic processing of the customer invoice or vendor invoice, b. creation of an internal payment file, c. clearing of intercompany accounts, d. internal statements for the subsidiaries, e. clearing and reconciliation of accounts and posting of internal statements, and f. periodic sweeping of intercompany accounts. Inhouse cash allows for the enterprise to concentrate its activities with fewer banks, allows transaction control over international borders, and concentrates cash resources and liquidity for multinational operations.

R/3 Cash and Liquidity Management allows for effective cash flow management over international borders. Cash and Liquidity Management is also in a central location and manages electronic payments and electronic banking, optimizes cash flow control, and automates and standardizes business processes including period—end reconciliation activities. The CLM process is as follows: a. Daily bank statement delivery, b. processing of inbound data, c. analysis and decision making on the cash forecast, d. borrowing and investment through the Treasury department, e. processing of outbound data, f. monitoring and control, followed by g. payment order processing. This process enables efficient communication with financial institutions and streamlined processing of financial information, global integration of information systems using ALE, better liquidity management, a central view of the liquidity situation of the enterprise, reduced borrowing costs and an increase in return on cash. CLM also allows the capture of market data for decision making, integration of global liquidity information, centralized and distributed Treasury operations, reduced transaction costs and hedging costs. CLM integrates with proprietary bank software, EDI or XML-based software; it allows for frequent bank postings and statement processing as well as lockbox. CLM also has electronic features for all banking transactions. Cash and Liquidity Management automates the banking process and eliminates redundant data entry, it enables thorough financial integration of the working capital accounts and bank accounts and ledgers are reconciled automatically. This helps to speed up the period—closing process, enables up-to-date cash reporting for the enterprise and therefore better decision making. CLM integrates with SD, MM, CFM and SEM and with banks and various bank financial platforms and has the ability to report on information in real-time.

mySAP Accounting Tools

SAP accounting has developed into an integral element of mySAP Business Suite that is geared toward an electronic solution to enterprise resource planning including an integrated interface that extends to the world wide web. This signals an irreversible transition from a more traditional enterprise resource planning solution to an extension of it including electronic based business. Included in this paradigm are companies striving for efficiency and growth on the one hand and greater control over finances and human resources on the other. SAP accounting will remain and integral part of R/3 Enterprise and mySAP Business Suite emphasizing simplicity of integration and easy upgrades that will enable the corporation to pick and choose along the path of upgrades and enhancements. MySAP Business Suite is a complete set of business applications for the international corporation. It has a stable, integrated platform enabling management to control the economic direction of the business and its networks more easily than ever and while maintaining competitiveness in an environment of change and adaptation.

FI remains at the core of mySAP Business Suite, based on SAP NetWeaver. SAP NetWeaver enables mySAP ERP to rest on an application platform including the web application server instead of R/3 Basis, as well as SAP portals, Business Information Warehouse, and other elements. R/3 Enterprise Financials include extensions in the areas of manual accruals, AIS, the automatic payment program, master data management, and enhancements to asset accounting among other changes. The biggest innovation to FI in R/3 Enterprise, however, is the Accrual Engine that is a subledger handling all contract-like information including payment history, etc. The accrual engine is a very flexible tool and can store postings, and process massive amounts of data in an optimized manner. The accrual engine enables automatic periodic calculation and posting of accruals that do not have to be fixed values, and the accrual engine is self-correcting. It supports multiple accounting systems and has a comprehensive information system that is integrated with Business Information Warehouse.

Asset accounting has new account assignment features where additional objects can be assigned for items such as real estate and public sector objects. The new assignments are available in the asset master records where the end-user can activate the account assignment objects and specify account assignment types for the different objects. Account assignment objects are company code dependent and transaction type dependent. There are also new check tools to prevent errors and problems in the assets portion of the IMG. The depreciation program has been redesigned and it is now possible to post depreciation to new account assignment objects, batch input is no longer necessary, and depreciation is now integrated with Schedule Manager where it has an error log, output lists and a job log. In the case of errors, the depreciation program (RAPOST2000) now has enhanced restart functions and the program can now post to multiple company codes. Asset Explorer now offers the possibility to display asset main numbers along with the asset sub numbers. Asset Explorer also allows transaction simulation and simulation for changes in depreciation terms.

In the Audit Information System, the reporting tree structure now emulates the structure of the balance sheet and evaluations with default variants have been created for every type of auditing function. In the automatic payment program, R/3 now performs an automatic analysis of the payment proposal and blocking of payments. This enables automatic settlement of the payment process, eliminates the need for a legacy system and is fully integrated with Schedule Manager. Other extensions in R/3 Enterprise Financials include a document relationship browser (answers the question: which documents relate to a business transaction or process), FI integration with CRM, enhanced exchange rate maintenance in the IMG, XML features in the tax application, enhancements to the cash journal, a DME engine (for incoming files), enhancements to Schedule Manager and Real Estate Management and others.

For more information on mySAP FI functionality and auditing in FI, contact http://www.fed-tax.com.

Some transaction codes and menu paths in the IMG and with SAP Easy Access

1. Account determination rule definition and assignment for assets—1. Define the rule: Financial accounting > Asset accounting > Organizational structure > Asset classes > Specify account determination; 2. Assign accounts to the rule: Financial accounting > Asset accounting > Integration with the general ledger > Assign general ledger accounts > Select chart of accounts.

2. Activate a depreciation type in each depreciation area—1. Ordinary depreciation—Financial accounting > asset accounting > Depreciation > Ordinary depreciation > Determine depreciation areas; 2. Unplanned depreciation—Financial accounting > Asset accounting > Depreciation > Unplanned depreciation > Determine depreciation areas.

3. Activate and configure some features of plants abroad—Transactions OBY6 and FTXP.

4. Activity allocation reposting—Transaction KB65.

5. Allow negative postings—Financial accounting > General ledger accounting > Business transactions > Document reversal > Permit negative posting.

6. Allow payments with payment cards—Financial accounting > Accounts receivable and accounts payable > Business transactions > Payments with payment cards > Make central settings for payments cards or assign a general ledger accounts to the cash clearing account.

7. Asset explorer—Transaction AW01N.

8. Assign a company code to a chart of accounts—Financial accounting > General ledger accounting > GL Accounts > Master records > Preparations > Assign company code to chart of accounts.

9. Assign a company code to a chart of depreciation—Financial accounting > Asset Accounting > Organizational structure > Assign chart of depreciation to company code.

10. Assign a dunning company code to various company codes (correspondence company code, too)—Transaction OBVU.

11. Assign a house bank and lockbox number to your house bank lockbox—Transaction OB10.

12. Assign accounts for general ledger account clearing differences—Financial accounting > General ledger accounting > Business transactions > Open item clearing > Clearing differences > Create accounts for clearing differences.

13. Assign currency and valuation profile to a controlling area—Transaction OKKP.

14. Assign depreciation posting frequency—Financial accounting > Asset Accounting > Integration with general ledger > Post depreciation > Specify intervals.

15. Assign elements to an asset class—Financial accounting > Asset Accounting > Organizational structure > Asset classes > Define asset classes.

16. Assign posting period variant to a company code—Financial accounting > Financial accounting global settings > Documents > Posting periods > Assign variants to company code.

17. Assign tax calculation procedure to country code—Financial accounting > financial accounting global settings > Tax on sales and purchases > Basic settings > Assign country to calculation procedure.

18. Assign tolerance groups to end-users—Financial accounting > financial accounting global settings > Document > Line item > Assign users to tolerance groups (employees).

19. Change general ledger account name—Financial accounting > General ledger accounting > GL accounts > Master records > GL account creation > Change GL accounts collectively > Change account name.

20. Check currency codes—In the reference IMG: General Settings > Currencies > Check currency codes.

21. CO user settings information—Transaction RPCO.

22. Configure authorizations for document type and business areas—In the IMG: Financial accounting > Financial accounting global settings > Document > Document header > Check display authorization for document type.

23. Configure asset accounting using the expert view—In the IMG: Financial accounting > Asset accounting > Overview for experts.

24. Configure currency and valuation profile—Transaction 8KEM.

25. Configure negative postings—Transaction OBY6.

26. Configure reconciliation ledger—In the IMG: Controlling > Cost and revenue element accounting > Reconciliation ledger.

27. Configure the payment program—In the IMG: Financial accounting > Accounts receivable and accounts payable > Business transactions > Outgoing payments > Auto outgoing payments > Payment method-bank selection > Configure the payment program (go through the buttons from left to right).

28. Configure the payment program application—Accounting > Financial accounting > Accounts payable > periodic processing > Payments > Environment > Maintain configuration.

29. Configures the posting information for the lockbox—Transaction OBAX.

30. Copy an existing chart of accounts to a new chart of accounts (chart of accounts view only)—Financial accounting > General ledger accounting > GL Accounts > Master records > GL Account creation > Alternative methods > Copy G/L accounts > Copy chart of accounts.

31. Copy planning data—Transactions KP97, KP98 and KP9R.

32. Cost center collective processing—Transaction KS12.

33. Cost center reconciliation—Transaction KPSI.

34. Cost element maintenance (FI)—Transactions FS00, FSP0 and FSS0.

35. Create a bank chain for a business partner—Transaction FIBPU.

36. Create a cash journal—Financial accounting > Bank accounting > Business transactions > Cash journal > Set up cash journal.

37. Create a credit control area—Enterprise structure > Maintain structure > Definition > Financial accounting > Maintain credit control area.

38. Create a general ledger account to capture exchange rate differences from accounts payable—Transaction OBXQ.

39. Create an account for the cash journal—Financial accounting > Bank accounting > Business transactions > Cash journal > Create general ledger account for cash journal.

40. Create asset classes from general ledger accounts—Transaction ANKL.

41. Create chart of depreciation with reference—In the IMG: Financial accounting > Asset accounting > Organizational structure > Copy reference chart of depreciation > Copy reference chart of depreciation.

42. Create cost center groups—Transaction KSH1.

43. Create foreign currency translation ratios—Transaction OBBS.

44. Create number ranges for checks—Financial accounting > Accounts receivable and accounts payable > Business transactions > Outgoing payments > Automatic outgoing payments > Payment media > Check management > Define number ranges for checks.

45. Create number ranges for customer accounts—Financial accounting > Accounts receivable & accounts payable > Customer accounts > Master records > Preparations for creating customer master record > Create number ranges for customer accounts.

46. Create number ranges for vendor accounts—Financial accounts > Accounts receivable and accounts payable > Vendor accounts > Master records > Preparations for creating vendor master record > Create number ranges for vendor accounts.

47. Create, change and maintain general ledger master records—Transactions FS00, FSP0, and FSS0.

48. Cross-application time sheets—Transaction CAT2.

49. Customizing settings for automatic credit control—Transactions OVA8, and FCV3.

50. Data dictionary—SE11. Use this transaction to verify the data class of a table—click the technical settings button and look at the field for data class: APPL0 is master data, APPL1 is transaction data, and APPL2 is organization and customizing.

51. Data transfer workbench: Fixed assets—Transactions SXDA and AR01.

52. Date determination for asset value—In the IMG: Financial accounting > Asset accounting > Transactions > Specify how default asset value date is determined.

53. Define a business area—In the IMG: Enterprise Structure > Definition > Financial Accounting > Define Business Area.

54. Define a posting key for a general ledger enjoy transaction—Financial accounting > General ledger accounting > Business transactions > General ledger account posting > General ledger account posting enjoy > Define posting key for general ledger account posting.

55. Define accounts for cash discounts granted—Financial accounting > Accounts receivable and accounts payable > Business transactions > Incoming payables > Payable receipt global settings > Define accounts for cash discounts granted.

56. Define accounts for cash discounts taken—Financial accounting > Accounts receivable and accounts payable > Business transactions > Outgoing payments > Outgoing payments global settings > Define accounts for cash discounts taken.

57. Define accounts for overpayments and underpayments—Financial accounting > Accounts receivable and accounts payable > Business transactions > Incoming payment > Incoming payment global settings > Define accounts for overpayments and underpayments.

58. Define accounts for payment differences—Financial accounting > Accounts receivable and accounts payable > Business transactions > Incoming payments > Incoming payments global settings > Overpayment/underpayment > Define accounts for payment differences.

59. Define and assign asset number ranges—1. Define: Financial accounting > Asset accounting > Organizational structure > Asset classes > Define number range interval; 2. Assignment: Financial accounting > Asset accounting > Organizational structure > Specify number assignment across company codes.

60. Define asset classes—Financial accounting > Asset accounting > Organizational structure > Asset classes > Define asset classes.

61. Define automatic payment posting information—Financial accounting > Accounts receivable and accounts payable > Business transactions > Outgoing payments > Auto outgoing payments > Automatic postings > Prepare automatic posting for the payment program.

62. Define company code—In the IMG: Enterprise Structure > Definition > Financial Accounting > Define, copy...company code > Edit company code.

63. Define default depreciation terms—Financial accounting > Asset accounting > Valuation > Depreciation areas > Determine depreciation areas in asset class.

64. Define default document type and posting key—Financial accounting > Financial accounting global settings > Document > Default values for document processing > Define document type and posting key.

65. Define depreciation area currency—Financial accounting > Asset accounting > Valuation > Foreign currency > Define depreciation areas.

66. Define depreciation areas asset value management—Financial accounting > Asset accounting > Valuation > Depreciation area > Define depreciation areas.

67. Define document change rules—Financial accounting > Financial accounting global settings > Document > Document header > Define document change rules.

68. Define document number ranges—Financial accounting > Financial accounting global settings > Document number ranges > Overview.

69. Define document types for enjoy transactions—Financial accounting > General ledger accounting > Business transactions > General ledger account posting > General ledger account posting enjoy > Define document types for enjoy transaction.

70. Define fields to match for the automatic clearing program—Financial accounting > Accounts receivable and accounts payable > Business transactions > Open item clearing > Prepare automatic clearing.

71. Define financial accounting document types—Financial accounting > Financial accounting global settings > Document > Document header > Define document types (Overview).

72. Define general ledger accounts for exchange rate differences created during clearing and define clearing rules (for account types)—Transaction OB09.

73. Define general ledger accounts for foreign exchange rate clearings—Financial accounting > General ledger accounting > Business transactions > Open item clearing > Define accounts for exchange rate difference.

74. Define general ledger accounts for intercompany or cross company entries—Financial accounting > General ledger accounting > Business transactions > Prepare company code transactions.

75. Define general ledger tax accounts—Financial accounts > Financial accounting global settings > Tax on sales and purchases > Posting > Define tax accounts.

76. Define holdback rules for terms of payment—Financial accounting > Accounts receivable and accounts payable > Business transactions >

Income invoices-credit memo > Define terms of payment for hold-back-retainage.

77. Define house banks for the automatic payment program—Financial Accounting > Bank accounting > Bank accounts > Define house banks.

78. Define how depreciation areas post to the general ledger—Financial accounting > Asset Accounting > Integration with general ledger > Define how depreciation areas post to general ledger.

79. Define line layout for open item processing—Financial accounting > General ledger accounting > General ledger accounts > Line items > Open item processing > Define line layout.

80. Define location for fixed asset master records—Transaction OIAS.

81. Define maximum exchange rate difference per company code (foreign currency)—Financial accounting > Financial accounting global settings > Document > Document header > Maximum exchange rate difference > Define maximum exchange rate difference per company code (foreign currency).

82. Define parameters for low value assets—Financial accounting > Asset accounting > Valuation > Amount specifications > Specify maximum amount for low value asset.

83. Define payment block reasons—Financial accounting > Accounts receivable and accounts payable > Business transactions > Outgoing payments > Auto outgoing payments > Payment proposal processing > Check payment block reasons.

84. Define payment terms for customers and vendors—Financial accounting > Accounts receivable and accounts payable > Business transactions > Incoming invoices-credit memo or outgoing invoices-credit memo > Maintain terms of payment.

85. Define posting keys—Financial accounting > Financial accounting global settings > Document > Line item > Control > Define posting keys.

86. Define posting keys for clearing—Financial accounting > General ledger accounting > Business transactions > Open item clearing > Define posting key for clearing.

87. Define tax calculation procedure—Financial accounting > Financial accounting global settings > Tax on sales and purchases > Basic settings > Check calculation procedure > Procedures.

88. Define tax codes for sales and purchases—Financial accounting > Financial accounting global settings > Tax on sales and purchases > Calculation > Define taxes on sales-purchase codes.

89. Define the screen template for the account assignment model—Financial accounting > Financial accounting global settings > Document > Reference document > Account assignment models-Define entry screen.

90. Define the transfer of depreciation terms—Financial accounting > Asset accounting > Depreciation area > Specify transfer of depreciation terms.

91. Define tolerance groups for customers and vendors—Financial accounting > Accounts receivable and accounts payable > Business transactions > Incoming payments > Incoming payment global settings > Define tolerances (Customers).

92. Define tolerance groups for end-users—Financial accounting > financial account global settings > Document > Line item > Define tolerance groups for employees.

93. Define tolerance groups for general ledger accounts—Transaction OBA0.

94. Define translation ratios for currency translation—In the reference IMG: Global settings > Currencies > Define translation ratios for currency translation.

95. Define variants for open posting periods—Financial accounting > Financial accounting global settings > Document > Posting periods > Define variants for open posting periods.

96. Define void reason codes—Financial accounting > Accounts receivable and accounts payable > Business transactions > Outgoing pay-

ments > Automatic outgoing payments > Payment media > Check management > Define void reasons.

97. Distribute exchange rates—System > Services > Reporting > Report RFALEX00 Distributing exchange rates.

98. Document entry screen configuration for currencies—Transactions OB41 and OBC4.

99. Dunning configuration—Financial accounting > Accounts receivable and accounts payable > Business transactions > Dunning > Dunning procedure > Define dunning procedure.

100. Dunning notice form—Transaction FBMP.

101. Edit user parameters—Transaction FB00.

102. Enter a new exchange rate type—Transaction OB07.

103. Enter an invoice from an external vendor—Transaction FB60.

104. Enter cash journal transactions—Transaction FBJC.

105. Enter exchange rates—IMG General Settings > Currencies > Enter exchange rates.

106. Enter exchange rates—Transaction OB08.

107. Formula planning aids—Transactions KPT6 and KPPS.

108. General listing of most automatic account assignments—find this at transaction FBKP.

109. If you are using SAP for tax reporting: Define tax jurisdiction codes—Financial accounting > Financial accounting global settings > Tax on sales and purchase > Basic settings > Define tax jurisdiction code.

110. Import the lockbox file—Accounting > Treasury > Cash management > Incomings > Lockbox > Import.

111. Internal order layouts master data—In the IMG: Controlling > Internal orders > Order master data > Define order types.

112. Internal order line item reporting—Accounting > Controlling > Internal orders > Information system > Reports for internal orders > Line items.

113. Journal entry—Transaction FB50.

114. Line items for cost centers—Accounting > Controlling > Cost center accounting > Information system > Reports for cost center accounting > Line items.

115. Link alternative payor/payee in the business partner master records—Transaction FK01.

116. Lockbox configuration—a. Financial accounting > Bank accounting > Business transactions > Define lockbox accounts at house banks; b. Financial accounting > Bank accounting > Bank accounts > Business transactions > Payment transactions > Lockbox > Define control parameters; c. Financial accounting > Bank accounting > Bank accounts > Business transactions > Payment transactions > Lockbox > Define posting data.

117. Logical databases—SE36. To look for which logical database a table belongs to, choose Extras > Table usage.

118. Maintain company code global parameters—transaction OBY6.

119. Maintain cost center master data—Transactions KS01 and KS02.

120. Maintain cost center master data from the standard hierarchy—Transaction OKEON.

121. Maintain depreciation keys—Transaction AFAMA.

122. Maintain fast entry screens for general ledger account items—Financial accounting > Financial accounting global settings > Document > Line item > Maintain fast entry screens for general ledger account items.

123. Maintain field status variants—Financial accounting > Financial accounting global settings > Document > Line item > Controls > Maintain field status variants and groups.

124. Maintain fiscal year variant—In the SAP Reference IMG: Financial Accounting > Financial Accounting Global Settings > Fiscal Year > Maintain Fiscal Year Variant.

125. Maintain internal order settlement rules—Transactions KO01 and KO02.

126. Maintain planner profiles—In the IMG: Controlling > Cost center Accounting > Planning > Manual planning > Maintain user-defined planner profiles > Maintain user-defined planner profiles.

127. Maintain planner profiles—Transaction KP34.

128. Maintain the cost center standard hierarchy—Transaction OKEON.

129. Make changes to a chart of depreciation—Financial accounting > Asset Accounting > Organizational structure > Copy reference chart of depreciation > Delete depreciation area.

130. Name a chart of depreciation—Financial accounting > Asset accounting > Organizational structure > Copy reference chart of depreciation > Specify description of chart of depreciation.

131. Open and close posting periods—Financial accounting > financial accounting global settings > Document > Posting periods > Open and close posting periods; also from the general ledger, accounts receivable or accounts payable—Environment > current settings > open and close posting periods.

132. Periodic depreciation program execution—Transaction OBA7.

133. Periodic reposting segment adjustment—Transaction KSW5.

134. Plan cost center budgets—Transaction KPZ2.

135. Posting activity allocations—Transaction KB21N.

136. Posting manual cost allocations—Transaction KB15N.

137. Process lockbox transactions—Transactions FLB2, FLB1, and FOEBL1.

138. Reconcile vendor and customer intercompany open items—Transaction F.2E.

139. Report painter for cost center accounting—Transaction GRR1.

140. Request or plan or enter expenses from a business trip—Transaction TRIP.

141. Resource planning—Transaction KP04.

142. Run reconciliation ledger—Transaction KALC.

143. SAP Implementation Guide—Transaction SPRO. Many of the menu paths in this section are IMG menu paths.

144. Schedule Manager—Transaction SCMA.

145. Set tolerance group for general ledger account clearing—Financial accounting > General ledger accounting > Business transactions > Open item clearing > Clearing differences > Define tolerances for general ledger account.

146. Set up accounting transactions for the cash journal—Financial accounting > Bank accounting > Business transactions > Cash journal > Create, change, delete business transactions.

147. Summarize internal orders—Transactions KKRC and KKRA.

148. Variance reporting—Transaction KSS1.

Some SAP tables affected by accounting transactions

1. BKPF—Accounting document header transport table.

2. BNKA—Bank master record table.

3. BSAD—Secondary index for customer cleared items.

4. BSAK—Secondary index for vendor cleared items.

5. BSAS—Secondary index for general ledger cleared items.

6. BSEC—Cluster table for one—time account data.

7. BSEG—Cluster table for accounting documents.

8. BSET—Cluster table for tax documents.

9. BSID—Accounting secondary index for customers.

10. BSIK—Accounting secondary index for vendors.

11. BSIS—Accounting secondary index for general ledger items.

12. GLT0—Accounting master record debits and credits table.

13. KNA1—General data for customer master records.

14. KNAS—Customer master records table.

15. KNB1—Customer master records (company code) table.

16. KNB4—Customer payment history master data table.

17. KNB5—Customer dunning data (master data).

18. KNBK—Bank details master data for customers.

19. KNC1—transaction figures customer master data.

20. KNC3—Special general ledger transaction figures customer master data.

21. KNKA—Central customer master credit management data.

22. KNKK—Control area data for customer master credit management.

23. LFA1—Vendor master data (general).

24. LFAS—Vendor master data for VAT.

25. LFB1—Vendor master data (company code).

26. LFB5—Vendor master dunning data.

27. LFBK—Vendor master bank data.

28. LFC1—Vendor transaction figures master data.

29. LFC3—Special general ledger transaction figures for vendor master data.

30. NBNK—Bank number ranges table.

31. NKUK—Change documents number ranges.

32. PAYR—Payment transfer medium table.

33. PCEC—prenumbered checks table.

34. REGUA—User and time details for payment proposal changes.

35. REGUH—Payment program settlement data.

36. REGUP—Payment program processed items.

37. REGUS—Payment proposal blocked accounts.

38. REGUV—Payment program control records.

39. SKA1—General ledger accounts master.

40. SKAT—Chart of accounts descriptions table.

41. SKB1—Company code general ledger account master table.

42. T001—Company codes organizational table.

43. T001B—Posting periods table.

44. T005—Countries organizational table.

45. T005T—Country names organizational table.

46. T012—House banks table.

47. T012T—House bank account names table.

48. TBSL—Posting keys organizational table.

49. TBSLT—Posting key names.

50. TCURC—Currency codes organizational table.

51. TCURF—Conversion factors organizational table.

52. TCURR—Currency exchange rates organizational table.

53. TCURT—Currency code names.

54. TCURV—Exchange rate types for currency translation.

55. TCURW—Organizational table for exchange rate type usage.

56. TGSB—Business areas organizational table.

57. TGST—Business area names.

Chapter Three

Overview of mySAP
Reporting Tools

Most enterprise software reporting systems provide for printed copies of reports that usually run in batch after hours and consume much computing time and resources. Many of these reports are not available through the enterprise intranet, and do not have intuitive or interactive features. Some of the features offered by mySAP reporting not available in these legacy systems are: a. drilldown reporting, b. more stringent data integrity for reporting purposes, c. online access to reporting functionality, and d. online updating for reporting purposes. Reports generated in SAP can exactly duplicate the reports from the legacy system, and SAP contains thousands of standardized reports (see menu path Information Systems => General Report Selection; see also transaction SA38 and select the entries help button). Here are some helpful prefixes for reports you will be searching for at transaction SA38: a. Controlling—prefix J1, J5, J6; b. Accounts Payable—prefix RF; Personnel—prefix RP; Materials Management—prefix RM; Production Planning—prefix PP; Quality Management—prefix RQ; Sales and Distribution—prefix RM, RL, RV. Remember that every mySAP report is really a program that you execute or create, and you can use the transaction SARP to display report trees. You can use transaction SERP to create and change several SAP report trees, and with SERP only the structure nodes are visible in the tree. The reports (variants, and lists, etc.) assigned to a structure node are displayed as node contents when selecting the relevant node.

Most reporting functionality in SAP can be done with a simple reporting interface on the presentation level, and programming code is only used for very specific purposes for reports that otherwise cannot be generated. In an SAP installation most of the reports necessary, perhaps three quarters of the reports, to the proper functioning of the enterprise according to its strategic, operational and market goals, etc., are available through the standard reports. Reporting is also available through Business Information Warehouse, ABAP Queries, Logistics Information Systems, Report Writer and Report Painter, and these areas of SAP make up in total perhaps twenty percent of reporting needs. The final fraction of reporting needs in SAP is handled through custom ABAP reporting and third party software.

There is a method for assessing reporting requirements that includes answers to the following questions: a. what is the use of the report? b. how is the report disseminated? c. can this report be replaced or augmented by another similar report? d. has the report been updated over time and is the report still necessary in the new reporting system? e. can the report be combined with another report to simplify the reporting process or eliminate needless reporting? f. is this report available in the standard SAP reporting functionality? g. what is the decision-making process that considers the necessity of the report and is it current? There needs to be in the enterprise a procedure for answering these questions and documenting report generation, destination (who gets the report?), configuration of the report and reporting deadlines. Whenever a request for a report is generated, the enterprise needs to inquire whether or not there is a standard SAP report that satisfies its requirements, and someone must make a decision about what tools to use to generate the report, or at least recommend the use of specific software tools if necessary.

SAP systems include a reporting interface through ABAP Queries (start with SQ01), interactive reporting including drilldown capability (start with transaction KCCF), Logistics (start with MCT0), Report Writer and Report Painter (start with transaction FGRP) for the following applications: General ledger accounting, legal consolidation, accounts receivable and accounts payable, bank accounting, asset accounting, SPL, funds management and travel management. Reporting using these tools is also available in other functional areas, for example: Cash management and budgeting, treasury and debt management, and risk management including investment management. Overhead cost controlling, product costing and profitability analysis are also available using these tools in the Controlling functional area. Other application areas for reporting include Enterprise Controlling, Sales and Distribution, Materials Management, Quality Management, Plant Maintenance, Production Planning, Project Systems, Personnel, Payroll Accounting, Training and Events Management, reporting on Basis Components and others. Reporting with mySAP also includes features such as integration with spreadsheet programs and the requirements of legacy programs, end user functionality, reporting optimization and rapid report generation, drilldown and report modification capabilities, ease of use including drag and drop capability not to mention the plurality of standard reports and the wide functionality of the reporting tools.

SAP reports can be scheduled to run without the end user monitoring the generation of the report. For instance, the end user can schedule the report to run monthly or weekly, etc.; after the occurrence of an event or after business hours. It is recommended that you create your reports as an SAP variant to avoid having to re-enter selection criteria, when you want to schedule the running of the report, and when the report will have many users and multiple key figures, for instance. Reports in SAP can also be displayed graphically, can be saved and sent through e-mail (menu path—System => List => Send). It is also recommended that you save your reports as lists in SAP for future reference and/or for future data integrity checks.

The ABAP Query

The ABAP Query (transaction SQ01, or menu path—Information Systems => Ad hoc reports => ABAP Query) can be used in any module to create reports. It is a relatively easy tool to use and allows the end user to select the table fields desired, sorting order of the data and presentation format for the data. The ABAP query also allows for report calculations and selection of fields from tables in different SAP modules.

As you know, all the data in SAP is stored in databases of which there are certain pre-defined "Logical Databases". LDB's pre-arrange and group data from the underlying database to make reporting in SAP an easier task. For a schema of common logical databases, see the end of this paper.

The ABAP query can be global or client—specific. In other words, global ABAP queries are available in all clients on a single server, and SAP delivers a library of sample ABAP queries that are transported through use of the ABAP Workbench. Client—specific ABAP queries are available only on the client level. The client level is the standard area for creating and maintaining the ABAP queries that can be transported on this level using the upload and download features of transaction SQ02. The procedure for creating an ABAP query includes seven steps, of the following: 1. Use transaction SQ01 and name the query, 2. Pick the SAP functional groups having the fields you want (SAP will prompt you to do this), 3. Choose your database fields, 4. Add your chosen fields to the report selection screen, 5. Specify the order and output of your fields in the report, 6. Make specifications in the selection screen, and 7. Publish the report. Transactions involved in creating an ABAP query: a. Use SQ03 to first create reporting user groups (define), b. Use SQ03 to assign users to the user groups, c. Use transaction SQ02 to create user functional areas (menu path—Information Systems => Ad hoc reports => ABAP Query => Environment => Functional Areas; then type in a functional area name and select the create button), and include the logical database and the fields you want to make available (define), and d. Use transaction SQ02 to assign functional areas to user groups. Remember to give names and numbers to the functional groups: Selecting the plus sign next to field names in the logical database adds them to your functional groups. When you are done with the selection from the logical databases, choose Save and then Generate. Return to the Functional Areas initial screen and choose the "Assignment to User Groups" button, then select the user group and save. After the user groups have been created, etc., end users should be locked out of SQ02 and SQ03. When a new user is created in SAP, the user should be added to the appropriate user groups. It is also important to remember to try above all to use the SAP-delivered logical databases here in order to avoid the complication of table programming.

The ABAP Query has advance reporting features that can be accessed through the Basic List Line Structure screen (step 5 in the previous paragraph) including additional control levels and control text, list line output options, field output options and field templates, list header and graphics. Advanced reporting features can be accessed by choosing the Change query button in the SQ01 screen.

Other Queries

There are some variations on the ABAP Query in present SAP systems, including Ad Hoc Query, InfoSet Query and QuickViewer. The Ad Hoc Query has been developed specifically for use in the Personnel (HR) application in SAP and is a single selection screen utility that uses SAP delivered "InfoSets" and has a quick unformatted presentation. Users of the Ad Hoc Query usually can only create and execute reports. The InfoSet Query is accessed through SQ01 and has a single selection screen with cross-module capability. The InfoSet Query operates with SAP delivered Infosets and is often used to create reusable reports to be stored in the Global application area, and to create Ad Hoc reports (simple data lookups) that can be stored in the global application area or client area. While these queries are useful and their functionality somewhat self-explanatory, I will concentrate this portion of this paper on QuickViewer.

QuickViewer is a single selection screen tool and produces limited lists. Quickviewer displays easily recognizable field names to make the formulation of reports with Data Browser all the more easy. Fields in QuickViewer are drawn from a logical database, InfoSet, or a table join, and QuickViewer does not require configuration of the data source before running the report. The InfoSet queries and reports created for other queries can also be used and reused in QuickViewer in the production environment, and vice versa.

To start a QuickView, use transaction SQV1 or menu path System => Services => QuickViewer. Enter a name for the QuickView and select the Create button. This will bring up the define QuickViewer popup box where you can enter a descriptive title, select your data source (logical database, table, InfoSet, or table join) and enter comments. Please try to follow your enterprise's naming convention, or at least have some sort of naming convention for QuickViews at this point and they will be easier to locate and work with later on. Select Basis or Layout mode in the popup box (Basis is the most common format; Layout is more graphical) and choose the enter button. In the next screen select the output format from the "Export as" box (ABAP List Viewer, List display, MS Word, MS Excel, Graphics, Local file, Display as table, or ABC Analysis) and select the report fields from the "List field select" tab. The "Data Source" tab in this box allows you to view the details of the data source you selected earlier. If you have selected the output to ALV (ABAP List

Viewer is the default setting for QuickViewer), you can sort the data in the report ascending or descending, filter, total, view in MS Excel, download to MS Word, save to a local file, send the report in an e-mail, perform an ABC analysis, view SAP Graphics of the report, and save and display as a variant among other functions. Finally, select the execute button on the application toolbar and a selection screen will appear if you are using a logical database. Make your selections and execute the report.

When you are using an InfoSet as a data source for a QuickView, you can add tables and fields to the report, including calculated fields. In order to create an InfoSet, use transaction SQ01 and enter a name for the InfoSet, and select Create. SQ01 will also show any existing InfoSets to choose from. Selecting Create brings up a popup box where you must enter a description in the name field and choose a data source (table join, table, logical database, or ABAP program) and choose enter. This brings up an InfoSet maintenance screen that lists the tables of a selected logical database, or data characteristics of other selections. You can expand the field groups to data fields on the left of the screen by double clicking on a field group. Right click on the fields and select "add field to field group" on the right hand side of the screen. Try to add fields to the field groups that correspond to table names to speed data retrieval when the report is run. Save the InfoSet and click on the Data Browser back button to return to the InfoSet initial screen; then follow the same procedure for creating a QuickView as with a logical database (above). QuickViews are generally used for rapidly creating individualized reports, though you can convert a QuickView into an SAP Query by using transaction SQ01 and following menu path Queries => Convert QuickView. When the popup box appears, enter your QuickView to be converted and follow the screens.

Downloading Reports

Reports from ABAP Query and other functionality can typically be made available to Microsoft Office applications through the following menu path: System => List => Save => Local file. To download database tables, use transaction SE11 or SE16, then the menu path System => List => Save => Local file. Any SAP data on the screen can be downloaded to Microsoft applications using System => List => Save => Local file. Remember the downloaded data is no longer protected by SAP security features once saved into a Microsoft application. Downloaded data is no longer real-time and any changes made to the downloaded data will not be reflected in SAP. Downloads can take a number of forms: a. Microsoft Excel format using the Excel button in SAP, b. Unconverted data to import into an application, c. Rich text format (RTF) often used with word processing applications, and d. Spreadsheet delimited format for MS Excel. Once downloaded or imported into MS Excel, the data can be subtotaled and summarized or filtered, or charted and graphed. Remember also that MS Excel has Pivot Table functionality that allows reformatting, cross-tabulating and summarizing of data from SAP. Microsoft Word integrates well with downloaded RTF documents from SAP in order to facilitate mailings to customers, printing product labels, envelopes, barcodes, or even e-mail communications. This often calls for the use of one of MS Word's data merging capabilities. Fonts for barcodes used in MS Word are relatively inexpensive and can easily be used to create labels for vendors or customers. SAP data can also be downloaded to MS Access if first saved to a text file and then imported into access. MS Access supports the following formats: MS Excel, text, Dbase software, FoxPro, HTML, and ODBC. MS Access has a report wizard that will enable you to work with the data you download. Access also enables the integration of reporting with SAP data and legacy system data.

There is reporting functionality in ABAP that is beyond the scope of this paper. For information on reporting in ABAP and ABAP programming, see the ABAP Objects reference (Jacobitz and Keller). For information on security in SAP Reporting, see first, Authorizations Made Easy, a guide from SAP Labs.

Some Commonly Used mySAP Reports

1. Accounts receivable information system—Program RFDRRANZ

2. Backorder processing—Transaction CO06

3. Backorders—Program RVAUFRUE, also V.15

4. Blocked sales orders—Program RVSPERAU, also V.14

5. Cost Center summaries—Plan/Actual/Variances—Program J1SIPTEX

6. Cost Center summaries—Target/Actual/Variances—Program J1SISTQX

7. Cost Center summaries—Program J1SISTQX

8. Cross-company code reconciliation—Program J5AB1TQX

9. Customer balances in local currency—Program RFDSLD00

10. Customer open item analysis—a/r aging—Program RFDOPR10

11. Display actual line items for cost centers—Transaction KSB1

12. Display actual line items for internal orders—Transaction KOB1

13. Empty storage bins—Program RLLS0400, also LS04

14. Evaluation of capacity planning—Transaction CM01

15. FI/CO reconciliation in company code currency—Program J5AF1TQX

16. FI/CO reconciliation in group currency—Program J5AF2TQX

17. Financial statements (b/s and p/l)—Program RFBILA00, also F.01

18. General ledger account balances—Program RFSSLD00, also F.08

19. General Ledger Accounts List—Program RFXKVZ00

20. General ledger line items—Program RFSOPO00

21. HR EEO-1 report—Program RPSEEOU1, also 4.5

22. HR entries and withdrawals—Program RPLEAT00

23. HR headcount report—Program RPSDEV00

24. HR new hire report—Program RPLNHRU0, also 4.5

25. HR payroll journal—Transaction PC1U

26. Incoming order report (statistical)—Program RMCV0100, also MCT0

27. Incoming orders customers/materials—Program RVAUFEIN, also V.12

28. Incomplete sales documents—Program RVAUFERR, also V.02, V.01

29. Inspection lost without usage decision—Program RQEVAM30, also QVM3

30. Inspection lot selection—Program RQEEAL10, also QA33

31. Inspection lots without completion—Program RQEVAM10, also QVM1

32. Internal order list—balance displays—Program J6L00TQX

33. Internal order summary—Plan/Actual/Variance—Program J6O00TQX

34. Invoiced Sales Report (statistical)—Program RMCV0100, also MCTA, MCTC, MCTE, MCTI, MCTG

35. List display of purchase requisitions—Program RM06BA00, also ME5A

36. List displays of purchase orders—Program RM06EL00, also ME2M, ME2L

37. List of all or open billings (documents)—Transaction VF05

38. List of all or open deliveries (documents)—Transaction VL05

39. List of customer open items—Program RFDOPO00, also F.21

40. List of posting change notices—Transaction LU04

41. List of sales orders—Transaction VA05

42. List of vendor open items—Program RFKOPO00, also F.41

43. Material documents—Program RM07MMAT, also MB51

44. Material stock list—Program RLS10020, also LX02

45. Missing parts information system—Program PPCMP000, also CO24

46. MRP list—collective display—Transaction—MD06

47. Multi-level BOM—Transaction CS12

48. Order cost element display—Transaction IW32, IW33

49. PM list of notification tasks—Program RIQMEL30, also IWW66, IWW67

50. Production order overview—Program PPIOA000, also CO26

51. Shipping report (statistical)—Program RMCV0400, also MCTK

52. Shop floor system information—Transaction MCP0

53. Stock overview—Program RMMMBEST, also MMBE

54. Stock requirements list—Transaction MD04

55. Vendor balances in local currency—Program RFKSLD00, also F.42

56. WM stock overview—Program RLLS2600, also LS26

Here Are Some Sample mySAP Logical Databases

1. $$M   Processing without database

2. $$S   Processing without database

3. 50V   Delivery in process

4. AAV   Logical Database RV: Sales Documents

5. ABS   ABAP Book: Customer and bookings

6. ACS   Example database for reading archives

7. ADA   Assets database

8. AFI   Logical database for orders

9. AGF   Accruals/Deferrals

10. AKV   Logical Database RV: Sales Documents

11. ALV   Archiving Deliveries

12. ARV   Logical Database RV: Sales Documents

13. ASV   Request Screen for Summary Information

14. AUK   Settlement documents

15. AUW	Allocation table
16. AVV	Archive Shipping Units
17. B1L	Transfer requirements by number
18. BAF	BAV—Data collector
19. BAM	Purchase requisitions (general)
20. BBM	Archiving of purchase requisitions
21. BCD	BC log. database: business partner fax locations
22. BJF	Loans flow records with date restriction(YR)
23. BKK	Base planning objects
24. BKM	Purchase requisitions per account assignment
25. BMM	Documents for Number
26. BPF	Treasury business partner
27. BRF	Document database
28. BRM	Financial Accounting Documents
29. BTF	Loan portfolios and flows
30. BTM	Process order; print
31. BUD	LDB for loans master data, conditions, documents
32. C1F	Cash Budget Management
33. CDC	Document structure
34. CEC	Equipment BOM
35. CEK	Cost Centers—Line Items
36. CFK	Data pool for SAP EIS
37. CIK	Cost Centers—Actual Data
38. CKC	Sales order BOM

39. CKM	Material master
40. CMC	Material BOM
41. CPK	Cost Centers—Plan Data
42. CRC	Work centers
43. CRK	Cost Centers—Total
44. CRZ	Logical database for courses BC220/BC230
45. CSC	Standard BOM
46. CSR	Logical database for archiving BOMs
47. CTC	Functional location BOM
48. D$I	Processing without database
49. D$M	Processing without database
50. D$S	Processing without database
51. D$V	Processing without database
52. DBM	MRP Documents
53. DDF	Customer database
54. DPM	Planned Orders
55. DSF	Loan debit position
56. DVS	Logical database for archiving DMS data
57. DWF	Loan resubmission
58. EBM	Purchasing activities per requirement tracking
59. ECM	Purchasing documents per material class
60. EKM	Purchasing documents per account assignment
61. ELM	Purchasing documents per vendor
62. EMM	Purchasing documents for material

63. ENM	Purchasing documents for document number
64. EQI	Logical database (equipment)
65. ERM	Archiving of purchasing documents
66. ESM	Purchasing documents per collective number
67. EWM	Purchasing documents per supplying plant
68. F1F	Funds Management: Fund, Funds Centers
69. F1S	BC: Planned flights, flights and bookings
70. FDF	Cash management and forecast
71. FEF	Cash management—memo records
72. FRF	Drilldown Selection Screen
73. FSF	Cash management—summary records
74. GLG	FI-SL summary and line items
75. GLU3	Flexible *G/L*
76. I1L	Inventory data for storage bin
77. I2L	Warehouse quants for storage bin
78. I3L	Inventory documents
79. IBF	Real Estate logical database (Rental agreement)
80. IDF	Real Estate logical database
81. IFM	General purchasing info records
82. ILM	Archiving purchasing info records
83. IMA	Logical database for cap. investment programs
84. IMC	IM Summarization (not usable operationally)
85. IMM	Inventory documents for material
86. INM	Inventory documents

87. IOC	Shop floor control—order info system
88. IRM	Reorganization of inventory documents
89. K1V	Generating Conditions
90. KDF	Vendor database
91. KIV	Customer—Material Information
92. KKF	Balance audit trail of open items
93. KLF	Historical balance audit trail
94. KMV	SD Documents for Credit Limit
95. KOV	Selection of Condition Records
96. L1M	Stock movements for material
97. LMM	Stock movements for material
98. LNM	Stock movements
99. LO_ CHANGE_ MNMT	Logical database for engineering change management
100. MAF	Dataset for dunning notices
101. MIV	BC: Planned flights, flights and bookings
102. MRM	Reorganization of material documents
103. MSM	Material master
104. NTI	Logical Database Object Networking
105. ODC	Shop floor control—orders per MRP controller
106. ODK	Orders
107. OFC	Shop floor control—orders per prod. scheduler
108. OHC	Shop floor control—orders by numbers

109.OPC	Shop floor control—orders by material
110.PAK	CO-PA Segment Level and Line Items
111.PAP	Applicant master data
112.PCH	PD
113.PGQ	QM: Specs and results of the quality inspection
114.PMI	Structure database (plant maintenance)
115.PNI	PM planning database PNM Planning database
116.PNP	HR Master Data
117.POH	Production orders database—header
118.PSJ	Project system
119.PYF	Database for payment medium print programs
120.QAM	Inspection Catalogs: Selected Sets
121.QAQ	Inspection Catalogs: Selected Sets
122.QCM	Inspection Catalogs: Codes
123.QCQ	Inspection Catalogs: Codes
124.QEM	Result entry
125.QMI	Logical Database (PM Notifications)
126.QMQ	Inspection Characteristics
127.QNQ	Quality notifications
128.QTQ	Logical database for inspection methods
129.R0F	Archiving: Bank master data
130.R0L	Archive selection: Transfer orders (MM-WM)
131.R1F	Archiving: FI transaction figures
132.R1L	Archive selection: Transfer requirements (MM-WM)

133.R2L   Archive selection: Posting change notices (MM-WM)

134.R3L   Archive selection: Inventory documents (MM-WM)

135.R4L   Archive selection: Inventory histories (MM-WM)

136.RBL   Archiving of transfer requests

137.RHL   Archiving of inventory history

138.RIL   Archiving of inventory documents

139.RKM   Reservations for Account Assignment

140.RMM   Reservations for material RNM Reservations

141.RTL   Archiving of transfer orders

142.RUL   Archiving of transfer requests

143.S1L   Stock by storage bins

144.S2L   Warehouse quant for material

145.S3L   Stocks

146.SAK   Completely Cancelled Allocation Documents

147.SDF   G/L account database

148.SMI   Serial number management

149.T1L   Transfer orders by number

150.T2L   Transfer orders for material

151.T3L   Transfer orders for storage type

152.T4L   Transfer order for TO printing

153.T5L   Transfer orders for reference number

154.TAF   Treasury

155.TIF   Treasury Information System

156.TPI	Functional location logical database
157.U1S	User master reorganization: Password changes
158.U2S	User master reorganization: Password changes
159.U3S	User master reorganization: Password changes
160.U4S	User master reorganization: Password changes
161.V12L	Pricing reports
162.VAV	Logical Database RV: Sales Documents
163.VC1	List of Sales Activities
164.VC2	Generate Address List
165.VDF	Customer database with view of document index
166.VFV	Logical Database RV: Billing Documents
167.VLV	Logical Database for Deliveries
168.VPF	Interest/Repayment condition items
169.VXV	SD: Billing Document—Export
170.WAF	Securities position plus additional master data
171.WOI	Maintenance Item
172.WPI	Preventive maintenance plans
173.WTF	Securities positions and flows
174.WUF	Sec.-Determ. master data for positions
175._S	Processing without database

Chapter Four

An Overview of mySAP Strategic Enterprise Management

This chapter is intended for use as and informal introduction to the SAP Strategic Enterprise Management application for end users who have some experience in business financial planning and reporting and some SAP background including data warehousing. Some familiarity with SAP is assumed. For instance, there is no discussion in this text of logging on or off the system, navigation in R/3, basic SAP functionality and terminology, specific business processes, or why enterprises need integrated software. This paper is ideal for project team members, possibly even project managers, financial managers, controllers and corporate decision makers as well as individuals in investor relations of the enterprise.

This chapter does address the topic, though not exhaustively, of the advantages of using integrated software and the functionality of SEM components. It also addresses the particular SEM concepts and terms, and hopefully will aid the reader in understanding how SEM can support business processes, strategic decision making and strategic vision in the enterprise. The reader is cautioned that BW-SEM, while it is an innovative and information enhancing product that will add value to corporate performance, will not solve systemic problems within the enterprise. SEM is not to be construed as a "magic bullet" for planning, strategic or operational target setting, or creating what might be desired changes in strategic vision, operations, or corporate culture among other things.

At the end of your reading this chapter, you should be able to identify the different SEM components, know some basic SEM transactions, and understand the relationship between the SEM components and corporate decision making, strategic planning and strategic vision and how these contribute to increasing shareholder value. You will also understand the SAP process of establishing and measuring operational targets, benchmarking and reporting.

Strategic Management as it Was

In the past, competition in the marketplace emphasized the placement and geographic scope of products in domestic and international commerce. Strategic emphasis was placed on processes internal to the enterprise on as local a level as possible and as efficiently as possible. This process, known as the "inside-out" approach has become outmoded as it relies on the historical success of the business for innovation, it does not capture the true magnitude of external influences either on the market or the enterprise, and it rewards short-term financial performance as a measure of the corporation's overall success.

Because products and services have an increasing geographic and commercial scale and scope, competition in the market for products and services has increased greatly. Produce life cycles have shortened as well, due to a modern emphasis on innovation and change. More businesses are taking advantage of their ability to raise equity or debt in the capital markets where investors are more sophisticated and well-informed than they have been in the recent past. Human capital is more transferable, transportable, and mobile than it ever has been. People can choose their employers more freely and skilled labor is available from many places in the world—a change in the labor market that has become a reality, especially in western countries.

The Increased Role of Stakeholders in the Operations of the Enterprise

For shareholders, employees, customers, vendors, business partners and society, the economic landscape of the large enterprise has changed due to individual mobility, stakeholder activities, demand for real-time information, increased competition and the image of the enterprise in society. A corporation in operation today must constantly generate value for its stakeholders, especially its investors, in order to continue to survive. Businesses that continue to generate value for their stakeholders are in a virtuous circle of strategic, economic and political success. The arrival of any sort of expectations gap between stakeholders and the enterprise in the game of economic survival calls for new strategy and new approaches to managing the corporation.

Even with the right strategic approach, the economic and overall success of the enterprise is not guaranteed. Strategic implementation must be quicker than the competition, more reliable, and must add to stakeholder value. Challenges to stakeholders in the capital markets, at the board level and on the company level include: a. Attaining or maintaining an economic advantage, b. translating stakeholder expectations into business strategy, c. Communicating the strategic vision to employees, stockholders and the public, d. Effectively monitoring corporate performance, and e. Determining the business units in the enterprise that are the most profitable.

With the issues of stakeholder value in mind, it is clear that strategic planning and performance management have to be linked to the execution of any corporate strategy. This means establishing meaningful operational targets and communicating them to all employees of a corporation. Financial resources must also be made available to execute a strategy that must be monitored with tools that keep the enterprise acting on information and according to operational targets.

How SEM Does It

Strategic Enterprise Management, as a New Dimensions component of SAP R/3 links strategic planning and simulation with enterprise planning. Through the Balanced Scorecard and the Management Cockpit, along with other tools like Business Planning and Simulation, Knowledge Warehouse, measures, alerts and queries, SEM supports enterprise-wide strategy implementation from the planning process to research, feedback, execution and communication. SEM enables automatic sourcing of external and internal business information, speeds up legal and management consolidations, provides benchmarking analysis according to key performance indicators (KPI's) and interpretation models. It also integrates strategic, financial and operations information and links the enterprise management process to your most important stakeholders. One of the main advantages of SEM as it is delivered is its support of strategic management processes by linking strategy and operations, and better managing internal controls.

Through the implementation of SEM, an enterprise can create a vision to align everyone and everything in the organization behind strategic goals. SEM allows the monitoring of performance of any employee involved in strategic decision making. Every employee involved would establish operational targets with his/her manager within his/her managerial scope. These targets are entered into SEM and at any time an employee or manager can monitor performance through an electronic scorecard. The personal scorecard makes sure the employees in the corporate hierarchy stay informed about their performance and its contribution to corporate projects and goals. Communication about issues on the scorecard is available through video conferencing and e-mail.

The main focus of SEM are the economic drivers of shareholder value: Sales growth, cash margin, the cash tax rate, working capital, fixed capital, the cost of capital, and gaining competitive advantage. This is simply a variation on the method of discounted cash flow analysis of corporate business health, and SEM looks into the future to help decision makers choose between the most profitable ventures or projects. The seven economic drivers of SEM help the decision makers, and the entire enterprise to carry out the tasks that will result in competitive gain beginning with research and planning, target setting, communicating priorities to the business units and employees, and the adaptation of execution to the business strategy. This means improving the business processes that complement the strategic vision of the enterprise and that are likely to make the strategy happen. SEM signals a change from the report-based form of strategic planning to planning and forecasting and then to a dialog process where information and evaluation allow organizations to adapt to change more quickly.

SEM comprises five functional software components: Business Information and Collection, Business Planning and Simulation, Business Consolidation, Corporate Performance Monitor, and Stakeholder Relationship Management. The interaction of these functional areas enables the collection and timely processing of market and internal information that may or may not be coded; market modeling and simulation as it interacts with business strategy, enterprise planning and risk—adjusted value computations. SEM also facilitates legal consolidations, value adjustments and computation of economic profit under many scenarios. Performance is analyzed through the monitoring of the Balanced Scorecard and measurement of key performance indicators (KPI's), and processing of the information facilitates the decision-making process. Tactical targets can be changed. Enterprise structure can be changed, and stakeholders will know about the changes through SRM. This means strategic business planning at top levels leads to target setting that is then cascaded to the relevant business levels and units.

The SEM Environment—Business Information Warehouse

Business Information Warehouse is a powerful online analytical processing system, or what SAP calls a Decision Support Environment. BW uses integrated data that is historical and its performance requirements are not predictable. BW users do not have predictable system usage patterns. In addition, data in BW can be optimized based on performance characteristics, and hardware can be optimized according to the same criteria. BW is not a "real-time" system as it often makes use of transported, summarized data. Significant data history is required in order for BW to be effective and results as reported in BW are at a specific point in time. BW integrates with other SAP R/3 applications that furnish data that is often denormalized or restructured by BW for queries. Generally, in order for BW to be effective, the enterprise needs between 2–7 years of data, and data that is not necessarily associated with one application, but is integrated from various applications.

BW is generally used when there is a reporting need using ad-hoc analysis, with different levels of detail to be made available and to all types of users. Data sources include different platforms and applications as well as non-SAP information systems. BW can generally be described to have three layers—extractions, staging and storage, and presentation and analysis. Data extraction and transfer in BW is extensively automated and BW is a separate database from R/3 where metadata from the R/3 system is managed. Application data from various source systems can be staged and stored in multidimensional relational tables in BW. BW allows the end user to analyze data from R/3 applications or any other business application, including external data sources such as databases, online sources and the internet. BW is suited to processing very large volumes of operational and historical data, and the Business Content functionality of the system (optimized to handle core areas and processes) allows the end user to examine operational and business relationships everywhere in the enterprise.

The BW OLAP system interacts with OLTP applications using BAPI's. The extraction mechanisms in BW that prepare the data for presentation are already available in R/3. In non-SAP external applications, the data is extracted using third-party software. The central data repositories in BW are called infocubes. These software units form the basis for reporting and analysis in BW, and since BW queries always refer to one infocube each infocube should have its own dataset. Each infocube has objects attached to it like fact tables containing key figures, and several surrounding dimension tables. An infocube is assigned to an infoarea in BW and each one is supplied with data from an infosource. The browser for BW (Business Explorer) in the reporting process always refers to one infocube. Through a query tool, a set of infocube data is extracted based upon the selection of characteristics; you should always use a set that is not equal to the entire infocube in order to ensure optimum performance of reporting on the characteristic dimensions. Multidimensional analysis is also possible in BW in order to present a rapid analysis of multiple perspectives on the same data subset.

The central point of BW is the Administrator Workbench that maintains, controls and monitors the database (database design, maintenance and administration, data load monitoring, update monitoring, and scheduling and executing data loads). The administrator workbench is used in the design of BW components and customizing the database in addition to the other features mentioned above.

Integration of SEM with BW

SEM operates on top of SAP—BW and there are no consistency problems between the five SEM components and BW. There is no work to do to interface SEM with BW and SEM's OLAP structure is optimized for analytic tasks. Metadata and master data structures, characteristics and values and hierarchies from the ERP system can be copied into SEM, and metadata as well as application data are collected using BW standard procedures. The software architecture of SAP ensures that individual SEM components can use each other's functionality when required, and the SEM architecture is designed to enable access to all principal functions using a web browser.

All five SEM components operate on identical data in BW infocubes and their metadata and application data are completely integrated: They have identical dimensions, characteristics and hierarchies, key figures, charts of accounts, master data and transactional data. Measures in SEM can be defined in detail in an abstract business way independent from the technical data handling. Several derivations of the same measures can be defined and can be mapped to infoobjects in infocubes. Measures can be classified geographically, or by industry, etc. Measures in SEM are defined using the Measure Builder that is delivered with three KPI-based models: Balanced scorecard, Management cockpit, and the Measure Driver Trees.

SEM obtains its transactional data from BW, and if necessary, from external sources. Plan data in SEM is often input via data entry. The SEM database is created in BW and SEM can be installed on top of BW or as a separate instance from BW. Installing SEM on top of BW means that only one database has to be maintained (saving time, space and effort) and the BW database is always up to date with respect to SEM. If the enterprise has BW and wishes to run SEM separately, the SEM system does not add to the data warehouse, loads and locks in the data warehouse do not affect SEM and data changes in the SEM system do not directly affect the data marts of the data warehouse. If the enterprise wishes to implement SEM as a stand-alone application, the organizational unit installing SEM does not require a data warehouse (just an OLAP engine). If the SEM system is stand-alone, SAP delivers all BW functionality to operate SEM and get data from the ERP system, even from external sources.

Business Information Collection

Within SEM, Business Information Collection facilitates and automates collection of internal and external, qualitative and quantitative information. Traditional sources of information include local files, people, online databases, distribution services, and print media. A newer source of information that has vast potential in every business area as a source of information is the internet. On the other hand, the internet will only prove useful as a source if it is approached in a structured manner. BIC links internal and external data through retrieval and assignment to evaluation objects in BW. These characteristics and key figures can then be recalled later to support a management view or result.

BIC is composed of three SAP-delivered tools: The Source Profile Builder, the Editorial Workbench, and the Information Requirement Builder. SPB is used to maintain master data details for BIC sources; EWB searches the internet, pulls in data, edits that data and then links it to internal characteristics and key figures. IRB helps to automate searches by building profiles of information needs according to position, department, etc. and links the output from requirements to possible sources. Once the information has been collected and edited, it can be linked to the BW database and then accessed by SEM Business Planning and Simulation, Balanced Scorecard, and Management Cockpit. Once the information document is linked to characteristics and key figures, it is then stored in a BW document service (BDS).

The entire information collection process for the enterprise can be handled by BIC and the corporation can therefore avoid data deluges and other pitfalls of information collection. IRB facilitates initial streamlining of the information requirement process through providing templates for departments and positions. Once the request has been generated and received, the data can be collected, edited and linked to BW key figures and characteristics that are then stored in the BW business document service as infoobjects. Once this process has been completed, the document can be accessed through other areas in SEM (Balanced Scorecard and Management Cockpit). Documents from the BDS can be retrieved and sent to an end user through the use of e-mail.

Business Planning and Simulation

The traditional approach to business planning focused on a centralized department controlling the planning process, defining organizational and functional sections and the procedures for creating the sections as well as consolidations and the top–down, bottom–up iterative planning process. Premises and standards for planning were determined centrally as well as the timing and monitoring of the planning process. Plans were derived according to organizational units and consolidated and put through an iterative process to be finalized perhaps once a year.

In SAP BPS the organizational hierarchy is a matrix, and due to the complexity of the planning process, large numbers of staff from varying parts of the enterprise are involved in the planning process. This calls for the use of coordinating tools that streamline the planning process and provide for planning integration on all relevant levels. The BPS tools ensure that the plan is up to date, for rolling forecasts and other processes, reliable and available on time. BPS supports easy modeling of strategic scenarios and easy translations of strategy into operational target setting through allocation of resources and the planning of KPI's. BPS incorporates stakeholder expectations into planning processes and ensures consistent definition of planning methods and procedures throughout the global enterprise. BPS also assures enterprise-wide control and coordination of the planning process. BPS facilitates integration between organizational units and hierarchical levels, functional areas and planning and reporting processes; and allows easy distribution and implementation of the planning application with flexible and quick processes that can be started on demand without clinging to a static schedule. BPS also allows planning from a top–down or bottom–up basis and gives management the ability to link strategy with operational plans.

Implementing BPS enables the production of strategic planning at any level of the enterprise and across organizational units at any level of detail and into time horizons in the future. Standard planning tasks such as reporting or sales and profit planning can be implemented with a minimum of customization, and BPS integrates with market data. BPS planning methods are designed to be easy and flexible, regardless of planning area. BPS integrates with R/3, external systems, or as a stand-alone application.

BPS structures are multidimensional and allow scenario modeling, definition of planning areas and planning strategies on all levels. Planning functions include copying, allocations, revaluation, distribution (top–down), aggregation (bottom–up), trends and forecasts, and dynamic simulations. Planning interfaces allow for the use of Microsoft Excel, HTML, SAPGUI for HTML and Windows. Planning processes are controlled through workflow and a planning monitor. All planning data from BPS is stored in BW.

The data objects in BPS are: 1. The Planning Area—the business environment in which the planning takes place, an infocube in BW (accessed using a remote function call), defines the characteristics and key figures for planning and has several planning levels; 2. Planning Level—a set of characteristics, characteristic values, and key figures; selections such as organizational level and location are defined here where you restrict the amount of data supplied from the infocube subject to relevance to the planning task; you also determine here which planning functions are to be proposed; 3. Planning Package—a subset of a planning level or a selection of characteristic values out of a planning level.

Planning functions in SEM generate the plan data in conjunction with a determined parameter set. Planning functions hand all general techniques for generating or changing plan data and include the following features: Manual planning, dynamic simulation, copy—copy a data selection from one characteristic value to another (i.e., copy one plan version to another), delete—deletion of a data section like a plan version, distribution (top–down), revaluation—percentage revaluation (i.e., increase labor cost for a product line by a certain percent), and distribution—the total of a selected key figure over the data selection is distributed by key or reference data over the data selection (distribute total direct material costs according to planned sales volume). Planning functions also include forecasting—forecast a key figure on the basis of past values using statistical methods, formulas—selected key figures are computed on the basis of other key figures in the same data selection,—reposting—selected key figures in the data selection are moved from one characteristic to another, and exits to define customer specific planning functions. The parameter set or parameter group contains specific parameters for a function call and every planning function may have several parameter sets. The parameter set consists of the settings of the planning function including details on how data is to be changed when the function is executed, for instance (this is especially useful for revaluations). The planning function is carried out with one parameter set at a time and accesses the data of one planning package at a time. Planning functions and parameter sets or groups are defined at predetermined planning levels. The planning area is the highest-level object in BPS; the planning level is below the planning area. Subordinate to the planning level are the planning package and planning function and below that, the parameter set.

A planning package enables the end use to refine the setting of the planning level (it is a kind of subset of the planning level) and it illustrates the basis upon which the planning functions operate. Planning packages are kind of like work lists where tasks are defined under the larger rubric of the planning package and are processed under the planning package. Planning functions are used to enter and change planning data. Always choose a planning package before executing a planning function.

InfoCubes

In BPS, the infocube always needs to be defined as a transactional infocube. To set the transactions flag for an infocube that already exists: 1. If the infocube can be deleted—a. Go to infocube maintenance; b. Choose InfoCube > Delete Data content and set the flag for transactional; c. Save and activate the infocube. 2. If the infocube can not be deleted—a. Enter transaction SE16 with table RSDCUBE; b. Set the flag "TRASACT" to "X" for the two entries of the infocube (OBJVERS = "A" and "M"); c. In the case of an Oracle database, create "normal" instead of bitmap indexes. 3. Resetting the flag for an existing infocube (if there has been a mistake in setting up the infocube)—a. If the infocube data can be deleted, the procedure is the same as above (1.) as there is no possibility of keeping the data when the infocube is reset. The infocube process is one of data requests and functions. As soon as the number of records in the data request exceeds 50,000, for instance, the request is closed and rolled up in a defined aggregate. In BPS, rollup, definition of aggregates and compression are still possible.

Transactional infocubes are optimized to data handling as it is needed in BPS, but transactional infocubes can not be filled through BW data sources. There are some exceptions, but BPS data can only fill a table through an SEM transaction. Non-transactional infocubes are optimized for extraction. To use the actual data, create two infocubes with only one of them set as "transactional." There are also two options for using actual data in BPS: 1. Copy the extracted data from the non-transactional cube into the transactional infocube with a planning function using the multiplanning area; and 2. Create a multiplanning area over both cubes to use the actual data as reference data, and for reporting define a multicube in BW to bring the actual and planned data together. The design of different infocubes in BPS will differ with respect to indexing and partitioning. A multiplanning area is an alternative that deals with an actual infocube populated with extracted data, and a plan infocube populated by transactional data. Data is pushed into the multplanning area using planning functions. Multiplanning areas are necessary when the planning functions make reference to actual data in a basic, non-transactional infocube.

Infocubes and BW enable planning to be carried out on a combination of characteristics: financial statements, financial planning, cost centers, investment planning, sales volume planning, production planning, human resources, and procurement (a product and customer group report by region, for instance). The planning architecture, including infocubes, consists of a hierarchy of different levels at which the database can be delimited. The architecture also includes the planning functions that refer to the database and different levels of the hierarchy can have different business assignments such as region, department, division, or organizational units on different levels.

Other topics within BPS: a. Formula extensions allow additional computations on data in the data selection in a transactional infocube, including additional selection, conditions and functions; b. Top–down planning allows distribution of non-assigned values or the total value of a level to the levels below it using relevant key figures (sales, salaries, material costs, etc.) or a specific key (i.e., percentages); c. Seasonal distribution, with the aid of a "distribution key" allows values to be set over a specific time period, and total values can be distributed over individual periods of the planning period in total. d. Revaluation allows actual values for different key figures to be changed simultaneously by specifying percentage changes for them (the end user can use "object-dependent parameters to restrict the revaluation to a small area of the data structure); revaluation also allows selective recomputation for basic key figures.

The infocube is integral in enabling the multidimensional reporting capabilities of SEM—BCS. Characteristics determine the classification of datasets, including the consolidation unit, line item, and posting level, fiscal year, period, and so on, for example. Characteristic values are part of the BCS master data, and combinations of characteristics or the characteristic values themselves are often referred to as objects in documentation. There are also evaluation objects (a combination or relationship of characteristic values)—for example, a consolidation unit would be UK, an item would be Sales and the posting level would be Reported data. There are also many key figures that might be relevant to evaluation objects, and key figures in SEM—BCS are values from local and group currencies, and quantity computations based upon end-user developed formulas.

Infocubes are the base of the reporting and analysis architecture in SEM as reporting information is gathered from the appropriate infocube. Some of the more common reporting questions are among the following—a. how are my product orders doing in the top 10 countries? b. How have the pricing and delivery terms of my vendors changed over time? c. What is the source of the increase in lead time for a product? d. Will my sales and distribution support costs be in budget if I decide to hire more people? e. What is the relationship between sales growth and market growth? f. Why are my product costs higher in a certain region? Through use of the Business Explorer, reports can be defined and displayed. Business Explorer supports multidimensional drill down, and has data mining capability. Business Explorer is primarily based on Microsoft Excel extended by the BW OLAP engine functionality. A report in BE consists of one or several queries presented in workbook format—a query defines the data to select for the report in the Microsoft Excel workbook. Microsoft Excel has some advantages as a reporting tool: It is a well—known software package, it has powerful display and format capabilities, and has graphical, computational, and formula—driven enhancements, and can be used to create templates to ensure uniform report design. As another use for reporting, infocubes are used with Business Explorer in a geographical information system in SEM.

Business Explorer can also be used to define queries. Queries can be easily defined and assigned in the Business Explorer query definition screen. In this screen, query definition is handled completely by drag-and-drop capability. The structure of the infocube is displayed on the left of the screen in which all elements can be used for the query definition. These infoobjects (as well as predefind query templates) can be dragged into other areas of the screen for query structure (line and column), and to define characteristics for navigation purposes as well as the filter expression. The query structure is visible as it is being defined and execution of the query processes data and displays a result in a workbook that is configurable and enhanceable with Microsoft Excel functionality. A workbook can be saved in SEM or on the hard drive of the end-user PC. Queries can also be saved in the end-user favorites menu. The info-catalog is a database of the corporation's queries and is a tool for managing the queries. From the infocatalog, viewable in Business Explorer, you can copy queries to various catalogs and assign the queries to different users.

Data in the infocubes is analyzed by defining queries and executing them in Business Explorer. Due to multidimensional BW-SEM functionality, you can create different views on the data with different queries on the infocube. The OLAP engine manages the queries will taking data from the infocube and provides data navigation tools in order to create different views of the data. In the query process, Business Explorer requests the data from the query and presents a view of the stored data as only the data actually required is transferred from the infocube to the query. If an alternative view of the data is required by the end user, the OLAP processor gets the data again from the infocube and that data from the current and previous Business Explorer drill downs remains on the application server.

Modeling

Dynamic simulation can help the decision maker to understand operational cycles and the relationship between strategy and the effect of strategic decisions. Dynamic simulation helps to understand responses and counter—responses, the cascading effects of decisions, enables experimenting with delayed effects and comprehending complex information (both financial and non-financial). It also helps to understand how the future can be anticipated using historical information—what expected sales will be, what will be the effect in the marketplace of strategic decisions, what will give the enterprise a competitive advantage?

BPS also has applications that are SAP-delivered in addition to the customizable applications. The SAP-delivered applications have a standard user interface, though special interfaces can be developed, and they have a high level of flexibility. Example master data and transactional data are delivered with these applications with a pre-defined business model and logic, and all these applications are integrated. Investment Planning provides the ability to test the feasibility of investment opportunities and returns; Balance Sheet Planning allows business financial planning through financial statement scenarios; Profit Planning allows sales and profit planning using both top–down and bottom–up approaches; and the Capital Market Interpreter provides measures of shareholder value from the point of view of the corporation as well as from third-party stakeholders. These applications and the BPS interface in general (with Excel, HTML, SAP GUI, or dynpro programming) are designed to solve standard tasks with the least possible amount of customization and configuration. Applications can always be adjusted and customized for the specific enterprise.

Business Consolidation

The international corporation typically has complex capital and group structures that have to be managed effectively in an environment of decentralized management. Business units in local markets have to be managed locally in order to benefit more from their local economic potential. SEM has these issues incorporated into its structure—corporate as well as legal entity reporting, and local target-based management. SEM allows core competencies and competitive advantage to be identified more quickly and used more efficiently as new business concepts can be transferred rapidly from one group to another. As enterprise goals are frequently set from external stakeholders, such as the corporate shareholders, SEM enables better implementation of those goals internally. SEM allows thorough collection of business information and rapid dissemination of relevant information (monetary and non-monetary) to business groups within the global enterprise. SEM also unites the different aspects of legal and management accounting and the company is controlled according to internal processes that can be extremely detailed and flexible. The legal entity is no longer the basis of consolidation in all areas.

SEM uses the Consolidation monitor and various method builders in association with BW to reconcile internal and external group reporting: a. Legal and management consolidations; b. Accrual accounting versus management accounting (on a value added basis or ABC, for instance); c. With multiple data categories such as actual, data, forecast, and budget characteristics. SEM enables the automation of all consolidation processes, even complex ones including divestitures, transfers, and changes in capital interests, etc. SEM has the ability to consolidate in U.S. GAAP, IAS, or the German Commercial Code with a system of comprehensive reporting, rapid processing and easy handling of data. BCS also includes a system of authorizations as an internal control.

The consolidation steps in BCS are the following: 1. Create consolidation master data; 2. Select the basis for the financial statements (U.S. GAAP, IAS) and in what currencies; 3. Standardize the entries for the corporate-level valuation of data; 4. Currency translation; 5. Aggregate reporting; 6. Intercompany eliminations; 7. Debt and equity interest consolidation; 8. Reclassifications; 9. Final consolidation; and 10. Balance carryforward.

There are hierarchical displays in BCS including consolidation groups, financial statement items, a data monitor and consolidation monitor, information system, and parameter display. The end user can access all these functions through the BCS application menu, and organizational data is structured in hierarchies in the application menu from which you can maintain master data, starting and monitoring data collection and consolidation functions, and generate reports. Many of the complex functions in BCS (such as investment assignment methods and consolidations) use a hierarchical display with easy console navigation. A data monitor and consolidation monitor control the processes in BCS and cover the tasks required for the standardization and consolidation processes; they also allow for immediate execution of consolidation functions. Tasks in this "Status Monitor" are end-user defined, and are grouped according to their informational sequence (task groups are assigned to dimensions). The status monitor allows viewing of the overall status of consolidation groups and units, the detail status of each task to be performed, opening and closing of accounting periods. Tasks can be run in test or update mode and can be blocked or unblocked. The status monitor also features "traffic lights" for overall status, auditing information, filtering and updating to the next milestone in the consolidation process.

The smallest organizational unit in BCS is the consolidation unit where the data collection and processing takes place. Important attributes of this master data include description of the entity, address, data transfer method, financial data type, tax rate and local currency. Consolidation groups are time-dependent and version-dependent aggregations of consolidation units or other consolidation groups. Important attributes of the consolidation group are the group currency, consolidation frequency, period of the first consolidation, and definable divestiture accounting for all groups. User-defined dimensions represent different types of consolidations including at the company level, business area, profit center, and user-defined business units. Dimensions can be linked to each other through validation rules, consistency checks and in the reporting process.

The transactional data of all consolidation units and groups included in the consolidation must be uniformly systematized in order to create representative financial statements. To facilitate this, each consolidation unit is provided with one or more consolidation charts of accounts that structure the financial statement items in accordance with internal and external reporting requirements. In the consolidation chart of accounts, all financial statement items are arranged in logical array and in an item hierarchy. The consolidation chart of accounts is client dependent only and can be used in different dimensions. To allow data transfer in the consolidation process from transactional data to reporting a link must be created from each general ledger account and a secondary cost element as well as the consolidated financial statement items. This rollup method, the data identifiers in the transaction accounts are identical to the identifiers of the financial statement items in the process of periodic extracts. In order to ensure the identifiers are identical, the end user can create the consolidation chart of accounts from both the FI chart of accounts and the financial statement version.

Data in BCS is processed and validated. Data collection in SAP includes rollup from either FI or special purpose ledger, periodic extract from the general ledger, periodic extract through a step consolidation in FI, rollup from another BCS system, or planned extract from R/2. Data collection in a non-SAP system includes an upload using data extractors from non-SAP systems, or Microsoft Access data entry using an additional Microsoft Excel interface. Validation in BCS is user-defined and is also employed in other SAP applications. Validations take place in addition to the standard data checks in the BCS posting transaction—the end user can run validation for the following processes in BCS: a. Document header; b. Individual document line; and c. The complete document. Validation in BCS usually takes place after data collection and currency translation in the consolidation process. Validation across consolidation units and dimensions is vital for the integrity of legal and management reporting.

Other activities in BCS include intercompany eliminations, and investment consolidation. The intercompany eliminations is a process to make sure only payables and receivables from non-affiliate companies show on the balance sheet. Consolidation of investments is a process ruled by a hierarchical structure for reported data. Method definitions and method assignments are used in the consolidation of investments process that enables simultaneous and step consolidations for any corporate structure, support of multiple reporting regulations, automated amortization of goodwill and similar treatment of other investments and equity, automatic generation of posting documents, and processing of virtually all business transactions. Predefined methods are provided with BCS, and the end user will find definitions easily customizable through parameter selections. The allocation method uses inheritance logic in the case of multiple-level groups, and reporting is customizable, comprehensive, specific and unrestricted. BCS provides a reporting application accommodating power—user functionality for ad-hoc reporting, internal report processing, generation and publishing as well as company-standard reports.

Corporate Performance Monitor

SEM—CPM operates on top of the BW OLAP engine and has the following features: A presentation and GUI builder, interpretation and model builder, and performance indicator/key performance indicator builder. These features and the OLAP engine interact with visual templates, interpretation models, performance indicator/key performance indicator catalogs, and structural templates for activity/process models. The two major end user areas of CPM are the Management Cockpit and the Balanced Scorecard that are SAP-delivered with templates. The application enables integration of reporting on any scale, easy and flexible definition of reports, reporting on non-financial data, support for strategic decision making, support for value-based management, and a tool for translating strategic objectives into operational targets. CPS is an application that enables the enterprise to monitor and learn from review and change processes in a single-loop learning process, or double-loop learning process. The single-loop process is one of checking results and taking action. The double-loop process is one of strategic review, changing the focus of strategic goals and then reviewing the corporate mission. Close performance monitoring is the most efficient way to detect a gap in performance goals versus performance results. Variances can be analyzed with respect to source and cause as well as their effect upon corporate performance: a. In a single-loop process, analysis leads to feedback and action on the operational level. In this process the enterprise sticks to its strategy, but modifies the process of implementing the strategy. b. In the double-loop process variances are analyzed not only in view of operational feedback, but also considering the validity of a strategy and its planning and execution. The double-loop process will confirm or invalidate hunches or assumptions in a strategy, or feedback may indicate a strategic planning process must being all over again.

CPM operates with the BW OLAP engine, which can be the SEM instance itself, or can compliment a remote BW system that also hosts other data. The multidimensional OLAP analysis made possible by CPM and its presentation tool, the Business Explorer, includes tabular reports compatible for data manipulation by Microsoft Excel, drill-down capabilities, and data manipulation and mining capabilities. CPM has a number of tools of which the following—1. Business Explorer, enabling multidimensional analysis, table reports and BW business content such as financial statement reporting; 2. The Measure Catalog (hierarchical) provides measure description, data sourcing and mapping and other SEM business content that are the infrastructure for the CPM interpretation models; 3. Measure Trees that show mathematical measure definition and shows illustrations of operational drivers on relevant KPI's; 4. Balanced Scorecard approaches KPI visualization online or in a meeting room to boost strategic communication and effectiveness.

The Measure Builder is the infrastructure (logical) for all the KPI-based performance monitoring in CPM including the Balanced Scorecard, the Management Cockpit and Value Driver Trees. The measure catalogue is hierarchical and contains measure types such as financial (profitability, cash flow, rate of return, etc.) and logistic (cycle times, per minute measures, etc.). The Measure builder, delivered with a number of KPI's, takes an abstract textual description of a measure from a business perspective, independent of technical format, including business assignment and is populated with benchmarking information from R/3 or third-party vendors. Measure sourcing and mapping includes possible sourcing of infocubes to measures and definition of key figure computations based on the source infocubes. Measures can be mapped to infoobjects in different infocubes in SEM to define the sourcing of measure data and can be classified by industry.

The Balanced Scorecard

One of the difficulties of strategic decision making and implementation is the gap between the high-level formulation of strategic vision and the operational targeting and budgeting process. Often, strategic vision and the operational aspects of an enterprise are not closely integrated as management is judged by budget execution, the budget focuses on financial results and is not generated with strategic goals in mind, and there is a gap between the allocation of resources and initiative for the implementation of strategy. Balanced scorecard integrates top–down communication of strategic vision and implementation with the bottom–up activity of performance monitoring. The management process supported by BSC is dynamic and is not scheduled in a fixed time period. Setup of BSC starts with an analysis of the economic environment of the enterprise, and evaluation of current strategic parameters followed by the formulation of a detailed strategy. The corporate strategy is then deployed to all business units while taking into consideration the needs of the business units. Strategic deployment is assisted by employee training, human resources alignment, financial resources alignment, monitoring of business performance during implementation of the strategy, learning from variances between planning and implementation, and adjustment of the strategic plan. This process might take place as often as desired.

In BSC, strategy is broken down into a number of components called themes. Each strategy consists of a number of objectives assigned one of four perspectives—financial, customer, internal process, and learning and growth. Each objective is assigned measures to track performance, with targets that are goals associated with an objective stated in terms of its associated measures. Activities, call initiatives, are assigned to one or more objectives as well. Initiatives have a sponsor, a clear time period or deadline, a budget, and resources, etc. Objectives are linked in BSC on a cause and effect basis. BSC has a financial perspective as well as the following perspectives—customer, internal processes, learning and growth, and human resources, among others.

The Strategy Map, another component of BSC is a way to visualize the structure of a balanced scorecard: The scorecard perspectives are shown visually with their respective emphasis or focus, and each perspective contains initiatives, strategic objectives and assigned measures. The cause and effect links show the relationship between the perspectives and goals, etc. on the strategy map, and show how management expects the strategy to work in order to achieve its goals. Links are shown along with objectives and the status of the objectives in the strategy map. Perspectives make up the vertical axis of the strategy map, and the single strategies comprise the horizontal axis. Single strategies might be linked by cause and effect indicators or chains. A cause and effect analysis might show all objectives and measures, and their statuses in addition to indicating relationships between objectives. Cause and effect analysis is tabular and definable by the end user where the right side of the PC screen is a view of objectives and measures, or of objectives and initiatives.

Analysis views include a screen that shows the entire scorecard in one hierarchy, but the hierarchy can be changed to

a.   Perspective/objective/measure

b.   Strategy/perspective/objective/measure

c.   Strategy/perspective/objective/initiative

d.   Strategy/initiative

e.   Perspective/objective/initiative

f.   Perspective/measure, etc.

The right part of the analysis screen is for tabular analysis view of objectives and measures, or of objectives and initiatives, and the columnar structure of the tabular screen is end-user definable. Detail screens allow the viewing of different elements of BSC: 1. The objective detail screen offers information about an objective (including the definition of the objective, performance rating indicated by the status icon, assignment of the objective (automatic or by its owner), textual assessments of objective owners, and assigned measures and initiatives); and 2. The initiative detail screen show information about each selected initiative (including the initiative definition, performance rating indicated by the status icon, assignment of the initiative, assessment of the owner of the initiative, comments on the assessment by scorecard users, assigned objectives, expected benefits of the initiative, and budget and variances).

The measures detail screen shows information about the selected measure including measure definition (name, purpose, owner, formulation), performance rating indicated by the status icon, measure assignment (either automatic or by the owner), textual assessment of the owner, comments by other scorecard users, and measure graphs and time series. The measure driver tree is a screen where the user can choose part of a tree displayed on the main area of the screen (the tree looks like an organizational chart). A mouse—over function details the formulation of each measure and driver trees indicate how KPI's are influenced by other measures. The driver trees show quantitative and qualitative linkages (quantitative linkages allow sensitivity and what—if analyses, and qualitative linkages show the influence of measures on the KPI's). The driver trees of the measures detail screen support value-based management and provide visualization of driver maps for key value drivers. They also provide a visual presentation of value metrics, and EVA among other measures.

A multiple scorecard comparison captures two or more scorecards on the same screen over a time frame of up to twenty—five periods. Before selecting a comparison scorecard, you must have authorization to do so, and the two or more scorecards must have the same fiscal-year variants. A multiple scorecard comparison allows scorecard to be compared in end-user defined sets and scorecards can be compared even if the strategies are not exactly the same. The display for the scorecard comparison is hierarchical where all the objectives of the compared scorecards are presented, or only the objectives of the "leading" or benchmark scorecard can be made visible in the left column. Time series measures can be displayed, or other measures such as return on equity can be displayed for comparison either for a subset of scorecards or for all scorecards.

Feedback from objectives and measures allows BSC to make an interpretation of the corporate mission, strategy and operational targets. Strategy formulation in BSC has three basic issues about corporate strategic vision: 1. What are our core competencies; 2. Which products/services do we want to sell while relying on those competencies; 3. How do we want to target our products/services. In BSC, overall strategy is composed of objectives that are subordinate and structured by perspectives and by the overall strategy. The objectives are linked by cause and effect relationships determined by perspectives, and are assigned measures (KPI's). A reading of the measures is considered an indicator of the success of strategic objectives. BSC measures are not only used for reporting purposes, but also for target-setting for the overall strategy. On the highest level of all these tools is the corporate vision or a description of the goals of the enterprise used to formulate strategic vision and objectives. Below this are strategies to achieve the goals set by the vision or mission where measures are assigned (KPI's) along with strategic objectives and targets, and results are monitored. On the lowest level of this hierarchy is resource allocation and target-setting. Scorecards can measure the strategic goals and performance of the enterprise overall, strategic units, business units, products, business areas, markets, or any context measurable in strategic terms. Scorecards are time-dependent and only indicate results for a chosen period as determined by end-user menu choices.

The purpose of BSC is to measure the elements created and linked together on the strategic map within a context set by the end user. BSC offers different views for presentation of the strategic performance of the enterprise and the cause and effect relationships of key figures that might otherwise go unnoticed. Use of BSC relies on the elements created and customized that determine the form of presentation of measures, results and any evaluation of the overall or specific strategies. BSC also uses the same database or OLAP engine as some other SEM applications—BPS and BIC.

The SEM module gets data from several SEM components: 1. Business Information Collection (BIC) that gets data from financials and logistics, external data from market sources, and unstructured external data from press releases, articles, analysts' reports and competitor information; 2. Business Planning and Simulation that gets data from plan scenarios and forecasts; and 3. Business Consolidation that presents data by legal entity or internal entity (profit center), with or without eliminating adjustments. As part of the SEM database, BSC has access to all the data in the central SEM database and provides the ability to write measure back to the database through BPS.

Measures in BSC have a status and a score, and multiple measures can be assigned to a single objective to which a weighting factor can be added as opposed to computing for an average. Examples of statuses are—average, best case, worst case, weighted factor, formulae, and manual only. Computation of scores for higher BSC elements is available in SEM, and scores for objectives are based on weighted factors from measures and initiatives. Computation of scores for employees can be defined and integrated in HR. Computation of scores is very flexible and is aggregated from the measure level up to the score-card level.

Other issues with BSC: User settings can be personalized as well as presenta-tion and viewing (private views for one user and common views for all) of the data. Data entry screens can also be personalized and preferred views can be stored for any area for future reference (this is especially advantageous for the driver tree views). An end user can define a series of views and browse through them with business explorer.

Stakeholder Relationship Management

SRM is a tool designed to meet the corporate challenge of survival in the stakeholder network. It is formulated on the principle that all corporate value includes intangible assets such as level of brand awareness, size and scope of market reach and customer base, skill of the corporation's employees, size of the network of economic or business partners, etc. SRM provides communica-tion and knowledge to and from the marketplace in order to enhance share-holder value. It also monitors any perceived gap in enterprise value assessments by internal stakeholders and external stakeholders in order to reduce business risk and uncertainty of the enterprise and (again) increase shareholder value. Through SRM tools that enable consistent communication with stakeholders and increasing shareholder value, stakeholders will more likely approve of the enterprise on corporate governance issues. Communica-tion is the key in satisfying internal and external stakeholders and without it the corporation would suffer from increased uncertainty, risk, and perhaps faulty perceptions of stakeholders on governance issues. SRM uses a contact database, web portal, different means of communication, and support and analysis tools for internal and public relations.

The stakeholder master database contains professional details on people, data on objective structure (expectations), power structure (influence), risk struc-ture (what is at stake in the relationship), and notes and comments. The stakeholder database also has a contact database that manages information requirements of stakeholders and actions and initiatives concerning stakehold-ers. Queries on stakeholders, contacts, documents, information requests and expectations are SAP-delivered.

Management Cockpit

Management Cockpit was formulated to give various strategic views of the enterprise and has four primary areas: a. Key financial ratios and CSF's area, b. Market, customers and competitors area, c. Internal processes, productivity and quality improvement, and d. Strategic projects monitor. Each area has up to six logical views, each logical view contains up to six frames and each frame presents one KPI and two KPI's in relation to one another. Management Cockpit can be used as an online reporting tool with access to essential KPI information. The logical views for key financial ratios, etc. are Main, Finance, Market, Resources, Progress, and Critical Units. The logical views for Market and customers, etc. are Market Shares, Sales Activities, Economics, CSI, Team Base, and Competitor Watch. The logical views for internal processes, etc. are Cost contention, Productivity, Product Quality, Quality of the IT Department, Quality of the HR Department, and Plan. Monitoring Strategic projects has views for Actions and Projects. Management Cockpit supports management teams that are implementing strategy and need visual presentations and a communications center. This module can also be used at various management levels for monitoring strategic projects like a construction project or software implementation. Logical views are subordinate to the four principal areas and can be accessed by clicking on the proper area of the screen in the presentation of one of the four main areas. "Frames" are subordinate to the logical views and the data source for the frames is defined and assigned when Management Cockpit is set up. Frames have various characteristics. They may be tabular, a graph or chart (line, bar, or tachometer visual). Frames access different types of data through the use of remote function calls, especially qualitative data, and frames are customizable by the selection of characteristics for presentation. SEM modules BIC, BPS and BCS feed data to the central SEM database, and Management Cockpit accesses the central SEM database directly.

Budgeting with SEM

The traditional approach to capital and operational budgeting has been one of cost control and maximizing a rate of return. Cost controlling, is however, often the key aspect of budgeting or financial planning in the business sense and strict adherence to budgets has often been a source of reward for those operating in a stable business environment. The traditional model is not at stable when markets or economic circumstances are subject to rapid change and by its constraints can prevent responding quickly to economic developments, keeping talented personnel in the organization, promoting innovation in the business along with quality and operational excellence. Relationships with business partners can also break down when the traditional budgeting process does not consider the reality of economic changes in an organization. Traditional budgeting does not always motivate people to act in the best interest of the enterprise and in many instances, the traditional planning process is sequential and takes an inordinate amount of time (perhaps four to five months a year from start to finish) and management resources (managers often spend up to 20% of their time planning and budgeting). Traditional budgeting includes a process of extensive bottom–up planning, negotiations over fixed budgets, centralized decision making on resource allocation, a mountain of details, and annual planning with a backwards orientation to financial results. Through the budgeting process using SAP-SEM planning tools, budgeting can overcome traditional barriers by empowering quick decision making, collaboration in decision making, and building in adaptation to organizational and strategic changes to ensure the business maintains the correct strategic course. The basic SEM budgeting process comprises top–down and bottom up planning and feedback systems, agreed upon targets with external and relative measures with decentralized decision making, a forward-looking rolling forecast and a process that enables managers to focus more time on corporate performance, measures and initiatives instead of strict financial results.

The budgeting process with SEM enhances the power of decision making by emphasizing self-governance on the part of responsible personnel, encouraging responsible performance, creating a network structure for coordination and communication, and calling for ownership and leadership in the budgeting process. SEM budgeting also allows the setting of relative measures for evaluation, various strategic processes with anticipatory capabilities for the use of scarce resources. SEM budgeting also obviously has a different approach to measurement and control as well as the rewards for correct action and decision making.

SEM adopts relative rather than absolute targets and budgetary targets are based on KPI's as well as external factors. Under this scenario managers are able to build new initiatives from strategic goals rather than from localized

administrative concerns. SEM enables early warning communication to personnel about changes that might affect their business. SEM planning allows
decision making and initiative by people close to the market and provides
managers with the flexibility to act more propitiously to solve problems. Measures are timely communicated to all management levels simultaneously at the
level of detail desired by strategy. Communication of this sort includes leading
indicators, actual results and forecasts, among other tools.

Performance evaluation in SEM is based on relative measures that drive and
motivate innovation and improvement through the interaction of teams,
groups or companies. Strategic and operational planning are integrated with
top–down and bottom–up processes as well as a scorecard for applying the
strategy to projects and operational objectives. Value is not just driven by
financial measures as CSF's have both financial and non-financial value drivers, and scorecards can be cascaded through different levels of responsible personnel. Global budgets are enhanced and supplemented by detail budgets
instead of a group of detailed budgets for all areas. SEM has changed the
"inside-out" approach to one of external motivation, innovation, and initiative
for improvement. A dynamic reporting interface reflects the ability of the
enterprise to meet its specific goals with rolling forecasts that can be adjusted
for feedback from the enterprise or external market. SEM budgeting replaces
input cost element planning by activity-based planning at all levels including
resource requirements, cost element planning and area budgets—essentially an
output-based system.

SEM has the ability to provide responsible personnel with clearly defined
budgetary targets that are flexible and have associated KPI's and benchmarks
for reporting purposes. SEM can communicate strategic goals and objectives
to employees and allow them to take the initiative to meet established targets.
Rolling forecasts with leading indicators are possible and enable the initiative
and feedback processes while allowing managers to anticipate changes and
adapt their decision making. Flexibly managed targets can be established at a
high level in the hierarchy as well as detailed operational plans that can be
adjusted for the circumstances of the market and the enterprise. These goals
are nonetheless monitored to maintain alignment with established targets in
the integration of strategic and operational planning. Consistent and timely
data can be reported using multi-dimensional plan/actual data scenarios and
performance can be evaluated with multiple measures and at multiple levels

(comparative performance between groups is also an evaluation option) within the corporation. Both financial and non-financial value drivers support the evaluation process with support of activity-based planning instead of input-based planning moving toward the view of the corporation as an internal market. Strategic reviews might take place once or twice a year while performance evaluation is a quarterly ongoing process.

Strategic planning emphasizes evaluation, simulation, value computations and scenarios in SEM, while operational planning involves target setting from strategy to resource planning to forecasting. Target setting and forecasting are the integration points between strategic and operational planning processes that include strategic adjustments, evaluations and initiatives, performance measures, process optimization and detailed operational planning. Rolling forecasts in BSC enable the integration and feedback processes of target setting and the relationship between targets and operational plans.

Reporting in SEM provides flexible tools that register the corporation's ability to adapt to change, for instance. Graphical reporting dependencies are illustrated for operational and strategic KPI's; simulation allows for adjustment and recalculation of values and simulations can be stored in BW as queries. Drill-down reports provide the basis for partitioning strategy into objectives, measures, targets and action plans. Views of CSF's and organizational characteristics are also supported by drill-down reporting. This enables the allocation of resources according to strategy and gives managers, team members, etc. initiatives they can implement based upon expected benefits, budget and costs, status management and assessments and comments.

The SEM process, including the budgeting process, enables process improvements for the better utilization of corporate resources, better assignment of those resources and better relationships with stakeholders (i.e., customers).

SEM Implementation

The first real issue in an SEM implementation is whether the enterprise needs the BW-SEM system on a single server with SEM on top of BW, or with SEM on a separate server from BW running the OLAP engine. With SEM on top of BW the system needs less maintenance from a basis point of view because there are fewer servers to purchase and maintain. The versions of BW and SEM in this configuration are dependent and upgrades to BW and SEM must take place at the same time without exception. With SEM and BW on separate servers (the SEM Data Mart), the BW system can be upgraded on a different schedule from SEM, and this is helpful if operations and IT do not often get along. Under this scenario performance decreases slightly due to the remote function calls between the two BW systems, and extra servers are required to purchase and maintain.

Also, consideration must be given to determining how to choose the best SEM implementation process for the enterprise as the SEM process can involve significant effort expended to configure BW. Choosing the wrong implementation strategy means spending too much time or not enough time on SEM configuration, BW configuration, or both. Spending too much time means the SEM project will suffer delays and BW configuration and development costs will soar. Spending too little time means that rework increases, especially with BW, and the SEM project will again suffer delays. The right balance must be found between implementing quickly and eliminating the need for rework.

The order of implementation is important in addition to the process of implementation. Integrating BW and then SEM means there will be less rework once the SEM project begins-standards have been set and personnel have an understanding of BW design. On the other hand, the actual SEM implementation will be delayed with possible rework to BW as the SEM needs are unknown at the start of implementation. In addition, the project will usually call for BW technical staff implementing SEM functionality. Integration of SEM, then BW means that SEM will be implemented quickly and more BW requirements will be drawn out. On the other hand, more BW rework is guaranteed as the strategic BW design has not been developed before the SEM implementation.

Implementing both BW and SEM at the same time calls for the least amount of BW rework as the BW design and SEM needs are developed simultaneously—functional staff work on SEM and technical staff work on BW. Development and integration at the same time means there will be a slower implementation process overall, mainly due to developing the BW strategic design while the project is proceeding.

In the event SEM implementation is critical to operations, most data will be input into SEM-BPS, it is critical for SEM to be live in short order and the only use for SEM is BSC, then implement SEM first. BW implementations need to come first if SEM is replacing existing functional tools, or there are many data sources for SEM including legacy systems, and/or BW is to be the main reporting tool (management is unsure of the SEM functionality). A concurrent project might have considerations like SEM being critical to operations, BCS implementation is needed and BW is to be the main reporting tool. Management Cockpit should be implemented with a pilot project or a smaller rollout due to its popularity and questions about functionality. The corporation will also want to limit its KPI's to less than twenty and should be strategic because SEM could be overrun with request to build multiple driver trees for each KPI—so keep it simple. Management also needs to decide whether or not to implement R/3 SEM or BW-SEM depending upon your system and the role of BW in the enterprise. Also be wary of Microsoft Excel macros as ABAP can operate faster; also remember application design in BW and SEM correlates with performance.

While SEM is technically a BW add-on, it is its own stand-alone application and is a business application, not a technical application like a data warehouse. BW team members (technical) should be differentiated from SEM team members (functional)—they need to be different people with different skills. The configuration of business users, technical and consulting differs whether or not the SEM project is the initial implementation, or a follow-on project. If the SEM project is the initial one, fewer functional people need to be on the project team, and more consultants. In the follow-on project more business users need to be on the project team, and fewer consulting people with about the same number of IT people as under the initial SEM scenario. Use SEM consultants for SEM and BW consultants for BW; also remember that configuring SEM infocubes is not the same as implementing SEM.

Some tips: Find out first of all if the enterprise has a license for SEM as SEM is included in mySAP licenses. If you do not own SEM, then you need to buy an SEM and a BW license for each user. Then define your end-user group (small number, pilot project, or a larger number), but always consider a pilot project. Use a thorough planning and documentation process for the implementation project, and remember that the project can be delayed because customization plans were not specific enough for BPS, etc. Further, minimize custom coding as it is a burden on maintenance and a headache for your basis people. Decide whether or not the corporation's current strategic process is to be left alone or improved (an SEM project is an opportunity to improve on strategic planning and budgeting processes, and SEM is possibly the only application your executives will ever use), and consider using the applications as delivered as they are very flexible, are great templates for planning integration, and the project will be speeded up. If do not know about BSC, or have your own scorecards, find out about what is needed for scorecard content and start working on it before the project begins. If you already have BSC content, use the highest level of the enterprise as a business unit and build only those BSC's necessary for senior and executive management. In addition, scorecards should be set to automatic buffering and need to have consistent element dates in the implementation and should first be rolled out in a demo system. Make use of corporate performance monitor variables to reduce effort expended on SEM, and remember the measure tree is based on the BW query and can only be viewed one measure at a time.

Menu Paths

1.   Access the properties of Economic Profit at your enterprise—Strategic Enterprise Management > Corporate Performance Monitor > Strategy Management > Value-Based Management > Value Driver Tree > TASEM2 > Economic Profit.

2.   Consolidation Monitor—Business Consolidation > Consolidation > Monitor.

3.   Consolidation Monitor/Data Monitor—SAP Menu > Business Consolidation > Data Collection > Monitor.

4.   Create a Consolidation Unit—Business Consolidation > Master Data > Consolidation Groups > Hierarchy > Edit.

5.   Create a New Planning Profile—Strategic Enterprise Management > Business Planning & Simulation > Cross-Application Planning > Planning Profile.

6.   Display a consolidation chart of accounts master data—Business Consolidation > Master Data > FS Systems > Hierarchy > Edit.

7. Display a validation—Transaction SPRO, then SAP Menu: Tools > Business Engineer > Customizing SEM BCS; then SAP Reference IMG > Strategic Enterprise Management > Business Consolidation > Data > Validations for Reported Data and Standardized Data > Define Validations > Display Validation.

8. Display the Balanced Scorecard—Strategic Enterprise Management > Corporate Performance Monitor > Strategy Management > Balanced Scorecard.

9. Editorial Workbench for a research request—SAP Easy Access > Strategic Enterprise Management > Business Information Collection > Editorial Workbench.

10. Management Cockpit—Standard SAP Menu > Strategic Enterprise Management > Corporate Performance Monitor > Performance Measurement > Management Cockpit.

11. Planning Framework—Strategic Enterprise Management > Business Planning & Simulation > Cross Application Planning.

12. Retrieve a Document in SEM—Transaction OAER (then fill in your parameters)—SAP Easy Access SEM Menu: Tools > Business Documents Administration > Miscellaneous > Business Documents.

13. Source Profile Builder—SAP Easy Access > Strategic Enterprise Management > Business Information Collection > Source Profile Builder.

Chapter Five

An Afterword

This book is a brief compilation of the capabilities of the enterprise accounting and reporting function in mySAP systems. SAP is continually introducing new paradigms for enterprise applications, including accounting and its tools related to enterprise management. SAP systems offer an important opportunity for business innovation to expand its support of key processes in the evolution of information technology. This development is promising and inevitable.

SAP systems, especially mySAP FI, are leveraging productivity developments and the evolution of IT to improve business performance and further lead to the development of innovative applications and new software technology. Using best-of-breed applications such as mySAP FI and SEM, SAP systems will inevitably develop a more seamless and automated software using cross-functional processes and developing further the integration of collaborative and transactional features of these processes. The indirect goal of these developments is a reduction of the total cost of ownership and maintenance costs of enterprise software.

Future enterprise software development for mySAP FI and SEM will concentrate on developing functionality beyond simple processing and integration. Support of these systems will be the result of flexible and configurable processes enabled through collaborative efforts. Web-like features of mySAP FI and SEM will entail an increased aggregation of software processes using a comprehensive informational model that crosses traditional application boundaries. SAP systems in the future will provide increased functionality as a response to the increased expense of custom development, will leverage integration costs, and maintenance and product releases will be less chaotic for end-users.

How will this happen? SAP provides a standard user interface for FI and SEM, and in fact, for the functionality of all of its components that will be developed further with SAP Netweaver. SAP has a unified data repository package in Business Information Warehouse that draws from mySAP and legacy systems. BW will be extended further into web services, as well as the other modules, for instance, and the SAP enterprise information system will become even more integrated and distributed. This will call for the use of data mining, workflow, and other business processes in SAP to allow the kind of collaboration that is possible.

As a software package including mySAP FI and SEM, mySAP ERP will have a faster and higher return on investment; its implementations will be broken up into smaller projects calling for cross-functional optimization and platform flexibility. The lower total cost of ownership of mySAP ERP will give the enterprise a measurable strategic and cost advantage and a more flexible development environment for internal and web services designed according to the needs of the business. Interportability and connectivity will become increasingly important vis-a-vis customization. This will result in additional efficiencies in information delivery and analysis, and strategic decision making systems will be made faster, cheaper and more flexible. Execution will be easier. Serving the customer will be easier and at a higher level. Quality assurance will be easier, and the competitive environment will become increasingly more competitive. SAP systems as they develop will enable a significant strategic advantage to SAP customers, and will enable the business to more easily follow a vision of how to better produce its products and serve its customers.

Index

A

ABAP interpreter 7
ABAP List Viewer 60
ABAP Queries 56
ABAP Query 57
ABAP Workbench 58
Account determination rule 40
Account Groups 21
Account Types 21
Accounts receivable information system 61
Ad Hoc Query 59
ad-hoc analysis 77
Administrator Workbench 77
Advanced Planning Optimizer 22
Alternative currencies 12
Analysis views 92
application context 6
assessing reporting requirements 56
Asset Accounting 9
Asset accounting 39
Asset Explorer 20
Asset explorer 40
Assign a company code 41
Audit Information System 20, 39
authorizations 14
Automatic Payment Program 23

B

Balanced Scorecard 91, 103

BAPI's 77
best-of-breed 104
Biller Consolidator 36
Biller Direct 35
BPS structures 81
Budgeting 96
business application 6
Business Explorer 84
Business Information Collection 79
Business Information Warehouse 76
business planning 80
business processes in SAP 105

C

Cash and Liquidity Management 37
Change general ledger account name 42
Check currency codes 42
Client/Server 1
Closing periods in FI 24
communication 19
Company Code 25
Configure asset accounting 42
Consolidation 86
consolidation steps in BCS 87
Controlling Area 9
Corporate Performance Monitor 89
Create a credit control area 43
Create chart of depreciation 43
Credit Control Area 8

Credit Management 25, 37

Cross-company code reconciliation 62

customer requirements 3

Customization for sales and use tax 16

D

Data archiving 13

Data collection 88

Data dictionary 44

data in SAP 57

data mining 105

data objects in BPS 81

Define a posting key 44

Define company code 45

Define general ledger tax accounts 46

Define tolerance groups 48

dialog server 8

dialog structure. 7

dimensions 88

Document Type 26

drilldown capability 56

Dunning configuration 49

Dynamic simulation 85

E

Enter exchange rates 49

Enterprise Modeling 18

enterprise planning 74

enterprise software 55, 104

F

Field Status Groups 27

financial accounting module 9

Financial statements 62

Functional Areas 28

functional software components 76

functionality for VAT 16

G

General ledger account balances 62

General Ledger Posting 28

H

hierarchical display 87

I

implementation project 101

infocubes 78, 82

inside-out 73

inside-out approach 18

integrated software 72

integration points 3

intercompany eliminations 89

Intercompany Journal Entry 29

K

key figures 83

key performance indicators 75

L

legal accounting 5

license for SEM 101

List of customer open items 63

Lockbox 12

Lockbox configuration 50

M

Maintain field status variants 50

Management Cockpit 96

master record 5

Measure Builder 78

measure driver tree 93

message server 8

metadata 78

Microsoft Office applications 60

MS Access 61

MS Excel 61

multidimensional reporting capabilities 84

multilingual enterprises 3

multiplanning area 83

multiple scorecard comparison 93

mySAP Business Suite 38

mySAP ERP 105

mySAP Financials 5
mySAP Logical Databases 64
mySAP reporting 55
mySAP systems 104

N
NetWeaver 38

O
OLAP engine 78
OLAP system 77
older versions of SAP 14
Open and close posting periods 51
Open Items and Posting with clearing 29
order to cash 35

P
Performance evaluation in SEM 98
planning functions 83
Planning functions in SEM 81
Planning Profile 102
Posting Keys 30
posting periods 10
process automation 2
Profile Generator 15
purchase to pay 35

Q
QuickView 59
QuickViewer 59

R
R/3 Enterprise 3, 4, 20
rate of return 18
Reconciliation Account 31
Reconciliation ledger 11
Report Painter 56
Report Writer 11, 56
reporting and analysis architecture 84
reporting functionality 56
Reporting in SEM 99

requirements for reporting 11
Run reconciliation ledger 52

S
SAP Financials 3, 4
SAP functional groups 58
SAP Graphics 60
SAP Netweaver 105
SAP tables 52
scale and scope 73
Schedule Manager 13
seamless integration 4
SEM components 73, 94
SEM environment 18
SEM implementation 99
Server requests 2
shareholder value 75
Source Profile Builder 103
Special purpose ledger 10
Stakeholder Relationship Management 95
standardized reports 17
status monitor 87
Strategic Enterprise Management 72
Strategic implementation 74
strategic planning and simulation 18
supply chain management 34

T
Tax Jurisdiction 32
total cost of ownership 105
transaction dialog process 7
transaction processing 6
transactional infocube 82
Travel Manager 12

V
vendor master records 5

W
web services 105

978-0-595-35018-6
0-595-35018-6

www.ingramcontent.com/pod-product-compliance
Lightning Source LLC
Chambersburg PA
CBHW051252050326
40689CB00007B/1168